Ultrasonic Waves

Ultrasonic Waves

Editor

Jeff Winters

scitus
academics

Ultrasonic Waves
Edited by **Jeff Winters**

ISBN: 978-1-68117-659-8
Library of Congress Control Number: 2016946526

SCITUS Academics LLC,
616, Corporate Way, Suite 2, 4766,
Valley Cottage, NY 10989

www.scitusacademics.com

Notice

Reasonable efforts have been made to publish reliable data and views articulated in the chapters are those of the individual contributors, and not necessarily those of the editors or publishers. Editors or publishers are not responsible for the accuracy of the information in the published chapters or consequences of their use. The publisher believes no responsibility for any damage or grievance to the persons or property arising out of the use of any materials, instructions, methods or thoughts in the book. The editors and the publisher have attempted to trace the copyright holders of all material reproduced in this publication and apologize to copyright holders if permission has not been obtained. If any copyright holder has not been acknowledged, please write to us so we may rectify.

Printed in United States of America on Acid Free Paper

Preface

Ultrasonic waves are renowned for their wide-ranging applications. They can be employed in numerous fields such as medicine, engineering, physics, biology, materials etc. A characteristic presented in all applications is the easiness of the instrumentation involved, even knowing that the methods are frequently very complex, sometimes requiring analytical and numerical developments. Ultrasonic vibrations of frequencies are greater than the upper limit of the audible range for humans—that is, greater than about 20 kilohertz. A common use of ultrasound is in underwater range finding; this use is also called Sonar. Ultrasonic homogenizing is very efficient for the reduction of soft and hard particles. A wide range of intermediate and consumer products, such as cosmetics and skin lotions, pharmaceutical ointments, varnishes, paints and lubricants and fuels are based wholly or in part of emulsions. Emulsions are dispersions of two or more immiscible liquids. Highly intensive ultrasound supplies the power needed to disperse a liquid phase (dispersed phase) in small droplets in a second phase (continuous phase). Sonochemistry is the application of ultrasound to chemical reactions and processes.

Ultrasonic Waves presents a number of advanced applications of ultrasonic waves, presented by the researchers in their scientific fields world-wide. This book should be of valuable source of information for advanced graduate students, scholars, as well as for those employed in industry.

Table of Contents

CHAPTER 1

Application of Ultrasonic Waves for the Improvement of Water Treatment

Akira Hiratsuka[1], Dhundi Raj Pathak[2, 3]

[1] Department of Civil Engineering, Osaka Sangyo University, Osaka, Japan

[2] Engineering Study & Research Centre, Kathmandu, Nepal

[3] Solid Waste Management Technical Support Center, Ministry of Urban Development, Kathmandu, Nepal

ABSTRACT

Application of ultrasonic waves is one of the novel techniques for the improvement of water treatment process. In this study, ultrasonic waves were irradiated to drinking water for water softening process and other contaminants removal. The experimental results showed that this technique improved the water treatment process efficiently. The study also revealed that the various parameters such as amplitude, frequency and irradiation time could affect the efficiency of ultrasound techniques for the improvement of water quality.

KEYWORDS

Water Treatment; Ultrasonic Waves; Frequency; Irradiation Time; Water Quality

1. INTRODUCTION

Water is essential natural resources for the existence of life in the earth. However, sources of water are under crisis due to rapid urbanization and the increasing population, which demand more agricultural, industrial and domestic supplies. On the other hand, increasing anthropogenic activities in both developed and developing world have mounted heavy stresses on water quantity and quality [1].

An understanding that the water conservation at the source of drinking water rather than remediation and treatment is very important has been deepened. However, the water sources that have been already contaminated have to be treated to make safe drinking water. At present, the water purification treatment and its management have been fully carried out to respond citizens who require potable and tasty drinking water.

As a research theme of environmental engineering, numerous researchers are much interested in the development of new techniques to get a safe drinking water through purification of water resources [2]. But, development of a novel as well as sustainable and cost effective techniques for the removal of pollutants from drinking water is the challenging issue for environmental professionals.

The most widely applied water treatment processes, a combination of some or all of coagulation, flocculationsedimentation and filtration to reduce or eliminate turbidity and improve water quality [3]. Moreover, ion exchange, carbon adsorption, membrane processes and disinfection are generally used to remove contaminants in traditional water purification system. Above processes have some purification limitation and application problems such as high cost, ineffective for removal some pollutant, operation problems and generate toxic secondary pollutants [4]. Such a limitation can be eliminated by the application of innovative techniques such as semiconductor catalysts, forward osmoses, advances oxidation process and magnetic purification [5]. There are several applications of irradiation of ultrasonic waves into fluids [6,7]. It is reported that physical phenomena (effect) such as cleaning, emulsification, cavitation, promotion of reaction etc. happen when small amplitude and high-frequency waves (ultrasonic waves) are irradiated in the water with impurities. Ultrasound is cyclic sound pressure with a frequency greater than the upper limit of human hearing, which is approximately 20 kilohertz [8]. Among the numerous applications of ultrasound, the approach is used in the field of water treatment. The application of ultrasonic waves improves water treatment process efficiently.

Ultrasound waves were irradiated to water with impurities mainly to examine 1) a condition of ultrasonic waves to accelerate the reaction better and 2) a condition of the softenization of hard water and removal of other impurities from drinking water. Then the removal rate of impurities from drinking water has been shown when changing the amplitude and frequency of ultrasonic waves. Moreover, by changing the ultrasonic wave condition, fundamental investigation of the content of various water quality parameters such as calcium (Ca), sodium (Na), potassium (K), silicic acid (SiO_2), sulphate ion (SO_4^{2-}) and magnesium (Mg) etc. is conducted.

2. EXPERIMENTAL PROCEDURES

2.1. Ultrasonic System

An ultrasonic cleaner used in this experiment is shown in **Figure 1**. A specification of the ultrasonic cleaner is as follows: Model: SC-320F, Output: 300 W, Power source: AC100 V, 50/60 Hz 6 A, Oscillated frequency: 28 KHz ± 10%, Date of manufacture: May 2010, Equipment number: 3Y0101, Maker: STM Co. Ltd., Japan.

2.2. Experimental Condition

In this study, the standpoint is to understand how the hardness and other contaminants are decreased in each water sample by irradiating ultrasonic waves. An irradiation experiment was carried out using an ultrasonic cleaner as shown in **Figure 1** at a frequency of 28 KHz as mentioned above in this experiment. Here we confirmed that the influence on position of the samples is not relating to the result. An experimental condition is as follows:

Figure 1. Photograph showing ultrasonic system.

Irradiation time: 30 min., 20 min. and 10 min. (3 phases).

Test sample of drinking water focusing hardness: 6 kinds (**Table 1**).

It is easy for us to get all drinking waters except for drinking water 2 because of a purchasing system in which only special contracted buyer can get. Items of water qualities centering "hardness" such as Ca, Mg, pH and those regarding drinking water indexes for healthy (Ca and Na) and/or tasty (Ca, K, SiO_2, Mg and SO_4^{2-}) are picked up in **Table 2**. A theoretical calculation Equation (1) as mentioned below is well-known as an expression of hardness.

$$\text{Hardness } [\text{mg/L}] \cong (\text{Amount of Ca } [\text{mg/L}] \times 2.5) + (\text{Amount of Mg } [\text{mg/L}] \times 4.1) \quad (1)$$

A calculation result of "hardness" using Ca and Mg is almost as same as the value expressed in the label of each drinking water, respectively. However, regarding drinking water 2, although Ca and Mg were expressed in the label, "hardness" was not expressed. Therefore, Hardness has been calculated using the Equation (1). As a result, we got the value of 59 (mg/L) as a theoretical one, which is quite higher than that of actual value 15 (mg/L). In addition, this value 59 (mg/L) is quite higher in comparison with that of other 5 drinking waters. Then we tried to measure the amount of Ca and Mg regarding the drinking water 2 once more. As a result, we confirmed a difference between the value of the label and actual measured value. In addition, water qualities parameters such as Na, K, SO_4^{2-} and BOD at irradiation time 30 min are shown in **Table 2**.

2.3. Analytical Method of Water Quality

The analysis of the water quality was conducted in the laboratory. Simplified water quality analysis kit made by Kyoritsu chemical-check lab., corp., Japan was used for examining Hardness, pH, COD, Mg, Ca, SiO_2. and ICP emission spectrometry for Na and K, and ion chromatography for SO_4^{2-}, and diaphragm electrode method for BOD, respectively.

Table 1. Description of drinking water samples.

Soft water: (Hardness: less than 100)	• Drinking water 1 (Made in Europe)
	• Drinking water 2 (Made in USA)
	• Drinking water 3 (Made in Japan)
Hard water: (Hardness: 101 - 300) (Weak)	• Drinking water 4 (Made in Europe)
Hard water: (Hardness: 301 - 999) (Medium)	• Drinking water 5 (Made in Europe)
Hard water: (Hardness: more than 1000) (High)	• Drinking water 6 (Made in Europe)

3. RESULTS AND DISCUSSIONS

All measured results are shown in Tables 2-4. The pH value increases after irradiating in all cases. This estimates that sulfate (SO_4^{2-}) decreases and combines with a chemical compound (neutralization). That is, it is suggestive of the occurrence of the following reactions.

Table 2. Experimental results (Irradiation time: 30 min., water temp.: 22.7°C (start) - 27.9°C (end)).

Test sample ⇒ Water quality ⇓	Drinking water 1	Drinking water 2	Drinking water 3	Drinking water 4	Drinking water 5	Drinking water 6
Hardness (mg/L)	60 → 35	15 → 5	32 → 15	304 → 100	315 → 150	1468 → 150
Rr (%)	41.7	66.7	53	67.1	52.4	89.8
pH	7.0 → 7.7	6.7 → 8.3	8.3 → 8.5	7.2 → 7.8	7.8 → 8.2	7.4 → 7.8
COD (mg/L)	7 → 7	8 → 12	7 → 11	7 → 11.5	3 → 11	5 → 8
Rr (%)	0	−50	−57.1	−64.3	−266.7	−60
Mg^{2+} (mg/L)	8 → 8	12 → 1.5	1.9 → 0.5	26 → 15	20 → 15	74.5 → ≧20
Rr (%)	0	87.5	73.7	42.3	25	73.2
Ca^{2+} (mg/L)	11.5 → 11.5	3.9 → 1	9.6 → 7	80 → 30	94 → 30	468 → 45
Rr (%)	0	74.4	27.1	62.5	68.1	90.4
SiO_2 (mg/L)	35 → 35	3 → 0.8	90 → 90	15 → 15	10 → 10	8 → 8
Rr (%)	0	73.3	0	0	0	0
Na^+ (mg/L)	11.6 → 0.8	9.6 → 4.7	2.2 → 1.3	7 → 0.7	7.7 → 0.4	9.4 → 1.1
Rr (%)	93.1	51	40.9	90	94.8	88.3
K^+ (mg/L)	6.2 → 1.8	3.6 → 1.6	8.4 → 2.4	0.3 → 0.3	5 → 1.5	2.8 → 1.3
Rr (%)	71	55.6	71.4	0	70	53.6
TSO4 (mg/L)	7.5 → 24	7.0 → 19	3.0 → 25	12 → 33	100 → 110	1100 → 1100
Rr (%)	−220	−171	−733	−175	−10	0
BOD (mg/L)	3 → 3	4 → 4	3 → 3	3 → 4	2 → 6	2 → <2
Rr (%)	0	0	0	−33.3	−20	trace

$$Mg^{2+} + SO_4^{2-} \rightarrow MgSO_4$$
$$Ca^{2+} + SO_4^{2-} \rightarrow CaSO_4$$
$$2Na^{+} + SO_4^{2-} \rightarrow Na_2SO_4$$
$$2K^{+} + SO_4^{2-} \rightarrow K_2SO_4$$

Therefore these reactions seem to be roughly appropriate ($K^+ \fallingdotseq Na^+$, Ca^+, Mg^+) in considering the decreasing trend of the concentration. In addition, we see that SO_4^{2-} does not increase in spite of the decrease of Ca in drinking water 5 and 6. It is probably due to a problem of the analytical precision exists. Furthermore, the question has been raised on why SO_4^{2-} is originally high or where this ion comes from in drinking water 5 and 6. **Figure 2** shows a comparison of changing hardness between irradiation time 0 min. and 30 min (immediately after experiment). It is understandable that hardness in each sample is decreasing. First of all, drinking water 6 is about one tenth compared with at 0 min. and next, about one second for drinking waters 1, 3 and 5, and then, about one third for drinking waters 2 and 4, respectively. In the experiment, a decreasing of concentration of Ca and Mg is observed in the water. It is thought that this is due to an effect of ultrasonic waves. It is estimated that this is due to the liberation of the solid and solidification of the same substances (Ca, Mg) because suspended solids are observed in the water (concentration drop in water). And a slight increase of the temperature by activation of molecular motion due to the irradiation of ultrasonic waves is also observed. It is known from the standpoint of property of water that a saturated concentration increases associated with the increase of water temperature. However, the reverse change is brought in this experiment (decrease of saturated concentration). In addition, it is thought that a possibility of undergoing any oxidation-reduction reaction as the change for pH and COD can be seen.

Table 3. Experimental results (Irradiation time: 20 min., water temp.: 29.0°C (start) - 31.7°C (end)).

Test sample ⇒ / Water quality ⇓	Drinking water 1	Drinking water 2	Drinking water 3	Drinking water 4	Drinking water 5	Drinking water 6
Hardness (mg/L)	60 → 20	15 → 5	32 → 15	304 → 50	315 → 75	1468 → 75
Rr (%)	66.7	66.7	53.1	83.6	76.2	94.9
pH	7.0→7.2	6.7 → 6.2	8.3 → 7.5	7.2 → 6.8	7.8 → 7.2	7.4 → 7.2
COD (mg/L)	7 → 2.5	8 → 5	7 → 2.5	7 → 2.5	3 → 3	5 → 2.5
Rr (%)	64.3	37.5	64.3	64.3	0	50
Mg^{2+} (mg/L)	8 → 3.5	12 → 0.5	1.9 → 0.2	26 → 8	20 → 8	74.5 → 15
Rr (%)	56.3	95.8	89.5	69.2	60	79.9
Ca^{2+} (mg/L)	11.5 → 8	3.9 → 0.5	9.6 → 7	80 → 15	94 → 15	468 → 15
Rr (%)	30.4	87.2	27.1	81.3	84	96.8
SiO_2 (mg/L)	35 → 35	3 → 0.8	90 → 90	15 → 15	10 → 10	8 → 8
Rr (%)	0	73.3	0	0	0	0
BOD (mg/L)	3 → 3	4 → 4	3 → 3	3 → 2	2 → 3	2 → 4
Rr (%)	0	0	0	50	−50	−100

Rr: Removal rate.

Table 4. Experimental results (Irradiation time: 10 min., water temp.: 30.6°C (start) - 32.5°C (end)).

Test sample ⇒ / Water quality⇓	Drinking water 1	Drinking water 2	Drinking water 3	Drinking water 4	Drinking water 5	Drinking water 6
Hardness (mg/L)	60 → 20	15 → 5	32 → 15	304 → 35	315 → 75	1468 → 50
Rr (%)	66.7	66.7	53.1	88.5	76.2	96.6
pH	7.0 → 7.2	6.7 → 6.2	8.3 → 7.6	7.2 → 7.2	7.8 → 7.2	7.4 → 6.8
COD (mg/L)	7 → 1.5	8 → 2	7 → 2.5	7 → 2	3 → 2	5 → 2
Rr (%)	78.6	75	64.3	71.4	33.3	60
Mg^{2+} (mg/L)	8 → 3.5	12 → 0.3	1.9 → 0.1	26 → 88	20 → 8	74.5 → 8
Rr (%)	56.3	97.5	94.7	69.2	60	89.3
Ca^{2+} (mg/L)	11.5 → 7	3.9 → 0.2	9.6 → 7	80 → 15	94 → 15	468 → 1414
Rr (%)	39.1	94.9	27.1	43.8	84	97
SiO_2 (mg/L)	35 → 35	3 → 0.8	90 → 90	15 → 15	10 → 10	8 → 8
Rr (%)	0	73.3	0	0	0	0
BOD (mg/L)	3 → 3	4 → 4	3 → 3	3 → 6	2 → 5	2 → 4
Rr (%)	0	0	0	−100	−150	−100

Rr: Removal rate.

Figure 3 shows a comparison of hardness among irradiation time 30 min., 20 min. and 10 min. The experimental results confirmed an increase of about 5°C with respect to the irradiation time 30 min.

Then we carried out an examination decreasing irradiation time (20 min. & 10 min.) and recorded the corresponding temperature. In irradiation time of 20 min. and 10 min., the increase of the temperature came to about 2°C. Under this temperature, it was found lower value of many water quality parameters including hardness. It is observed in the experiment that the extent of suspended solid in the water after irradiating the time of 20 min. and 10 min. is almost equal to the case for 30 min. Therefore, it seems that the irradiation by ultrasonic waves is brought into the result in which many water quality parameters are shown lower tendency. However, at present it is very difficult to compare with the effect among irradiation times (30 min., 20 min. and 10 min.) quantitatively (for example, short irradiation time is more effective than long one regarding the change of "hardness."), which warranting further examination on this.

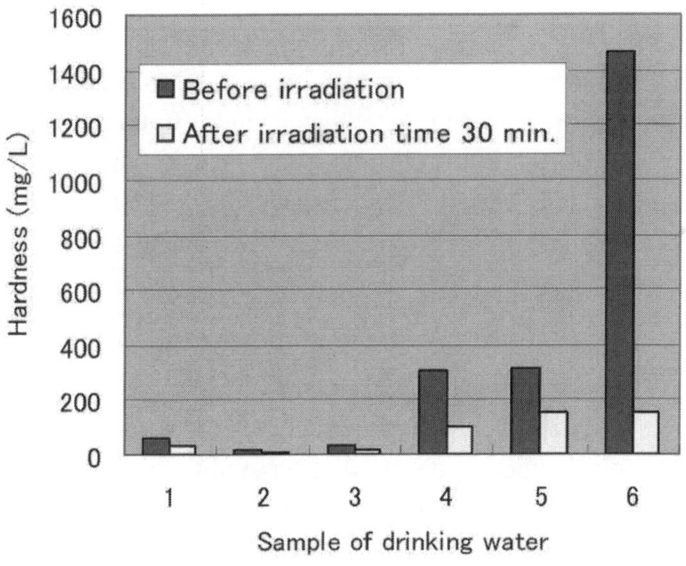

Figure 2. Change of hardness after irradiation time (30 min.).

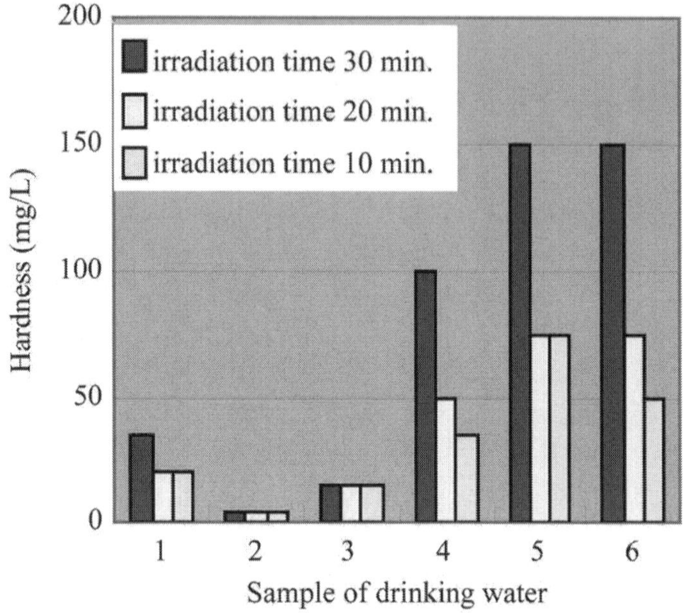

Figure 3. Comparison of hardness among irradiation time 30, 20 and 10 min.

Figures 4-8 show a change of water qualities such as Ca, K, SiO_2, Mg and SO_4^{2-} of drinking water at irradiation time 30 min. These items of water qualities are items on tasty water index (O Index) [9]. This O Index is shown in Equation (2), it is recognized that the value of more than 2 in O Index is tasty drinking water.

$$Ca + K + SiO_2 / Mg + SO_4^{2-} \quad (2)$$

In all drinking waters, there is not a large change in a denominator (Mg + SO_4^{2-}), but large change in a numerator (Ca + K [except for SiO_2]). Therefore, there seems to be a possibility that an irradiation by ultrasonic waves to drinking water increases the O-index which means that the taste of water is improved. **Figure 9** shows a change of water quality of Na of drinking water at irradiation time 30 min. This water quality parameter together with Ca are related to taste of drinking water which is called K index and given in Equation (3). It is recognized that the value of K index more than 5.2 is wholesome drinking water.

$$Ca - 0.87\ Na \quad (3)$$

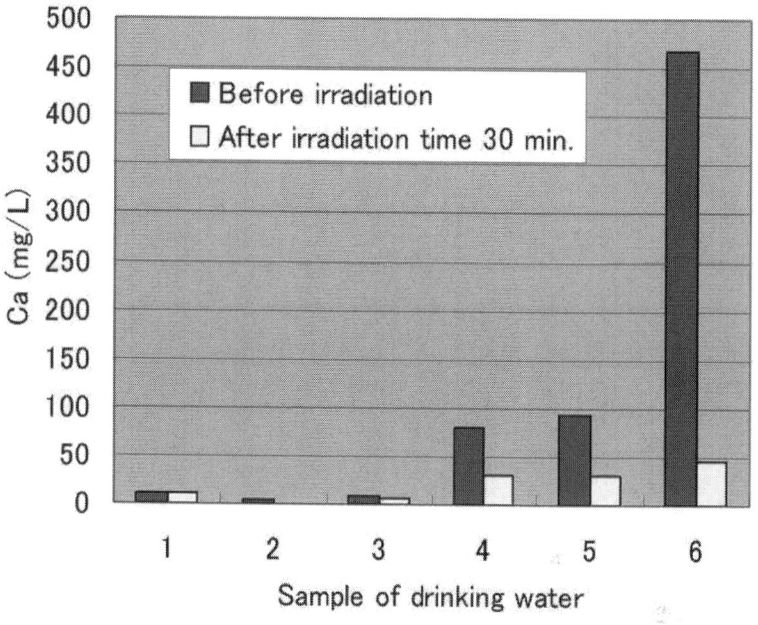

Figure 4. Change of Ca at irradiation time 30 min.

It does not exist large difference in the drinking waters 1, 2 and 3 for Ca, but exists large one in drinking waters 4, 5 and 6, respectively. In addition, there is a decreasing tendency for Na in all drinking waters. Especially, it is understandable that the drinking waters 2 and 3 have a large change. Therefore, there seems to be a possibility that an irradiation by ultrasonic waves improve the quality and taste of water significantly.

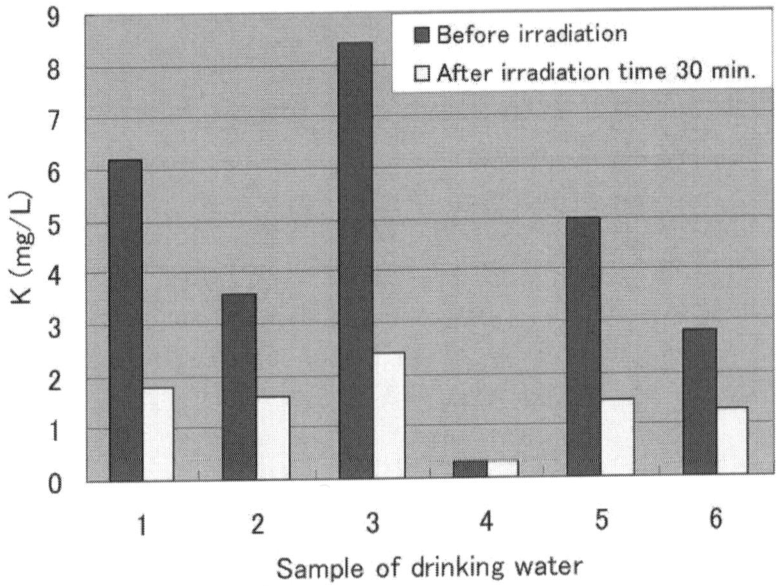

Figure 5. Change of K at irradiation time 30 min.

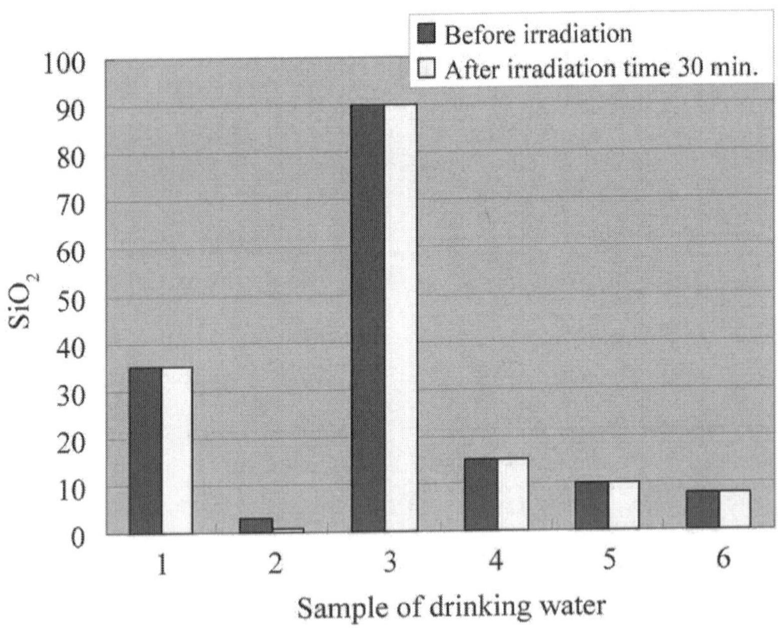

Figure 6. Change of SiO_2 at irradiation time 30 min.

Figures 10-12 show a removal rate of Hardness, BOD and COD in different irradiation time (10 min., 20 min. and 30 min.), respectively. As can be seen from Figures 10-12, the removal rate of the hardness component showed about 42% - 90% in irradiation of 30 min., about 53% - 95%, 20 min. and about 53% - 97%, 10 min., respectively. All the values showed the positive ones (in decrease). On the other hand, BOD and COD showed the negative ones (in increase) at approximately half of the samples. It can be seen from the change of the impurities (Tables 2-4) that the concentration of the impurities in drinking water decreases due to irradiation, and thereby the increase of the seeming rate of BOD and COD is considered. In that sense, it also can be thought that the impurities were removed completely.

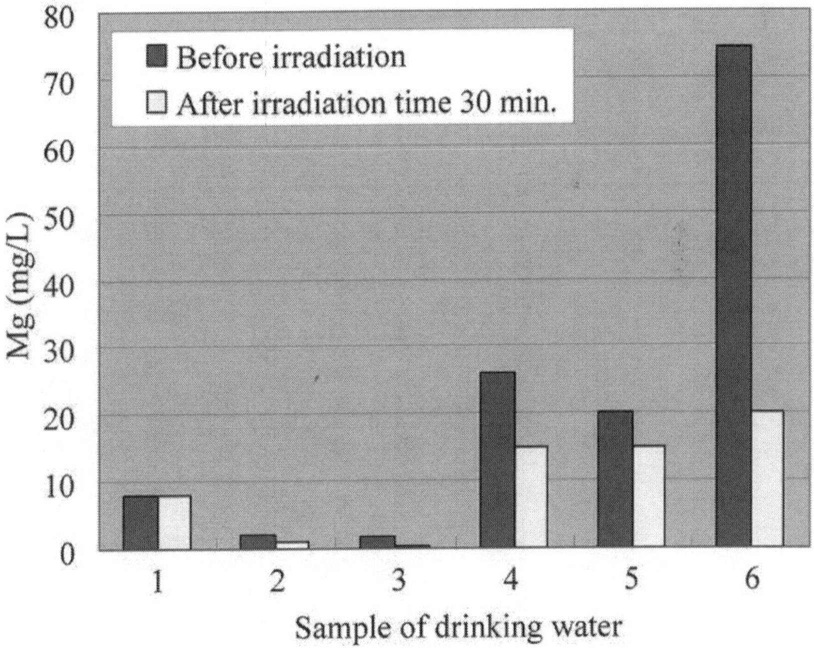

Figure 7. Change of Mg at irradiation time 30 min.

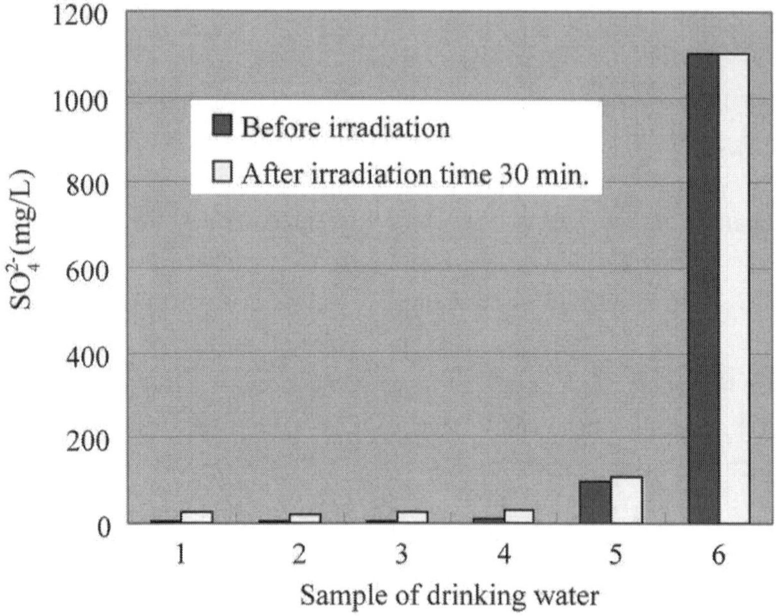

Figure 8. Change of SO_4^{2-} at irradiation time 30 min.

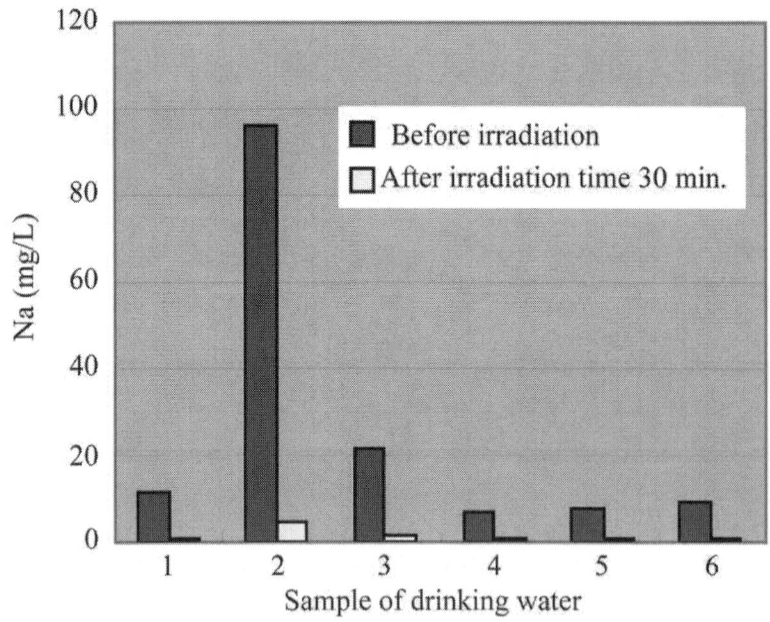

Figure 9. Change of Na at irradiation time 30 min.

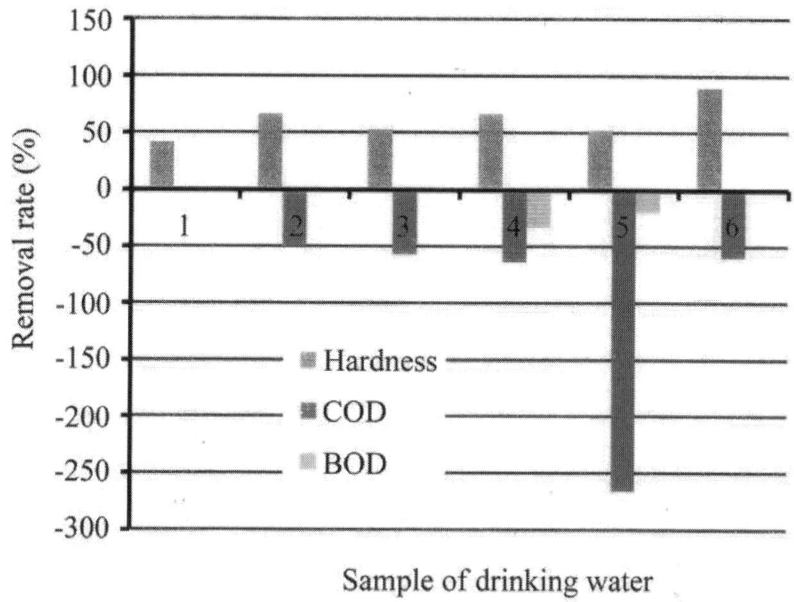

Figure 10. Removal rate of hardness, COD and BOD due to irradiation time (30 min.).

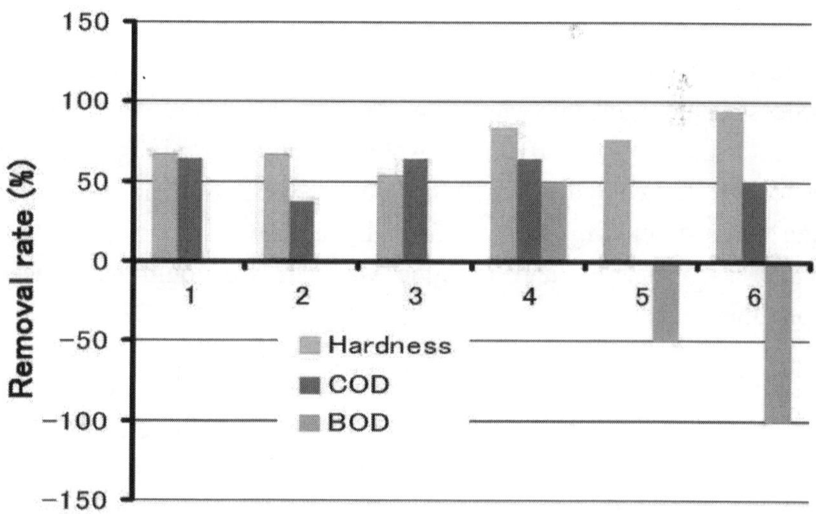

Figure 11. Removal rate of hardness, COD and BOD due to irradiation time (20 min.).

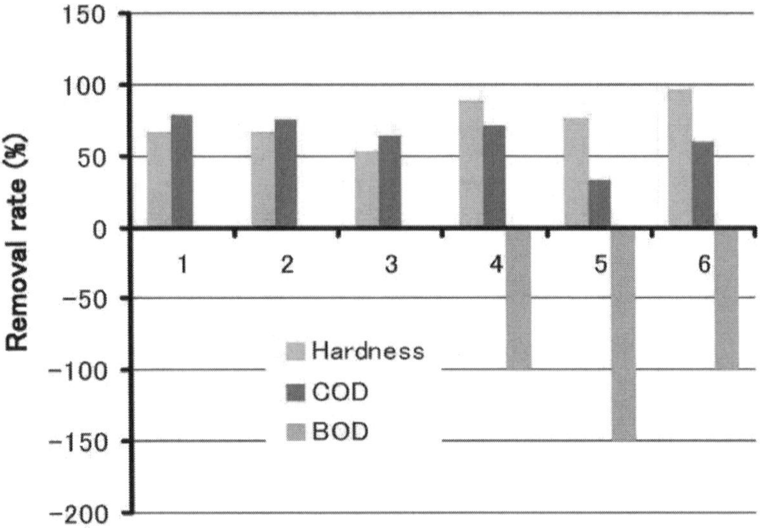

Figure 12. Removal rate of hardness, COD and BOD due to irradiation time (10 min.).

4. CONCLUSIONS

In this study, an examination for the improvement of water treatment process using ultrasonic waves was carried out. As a result, frequency of ultrasonic waves affecting water quality describing hardness, especially the content of calcium (Ca), sodium (Na), potassium (K) and magnesium (Mg) was confirmed. There will be a possibility to make tasty, healthy and immune water if we could have water containing many of the ingredients such as Ca, K and SiO_2.

Regarding the removal rate of the hardness, BOD and COD due to irradiation time, the removal rate of the hardness component showed high value of about 42% - 97% at each irradiation time. On the other, BOD and COD, the possibility of the complete removal of impurities is supposed at each irradiation time.

The experimental results confirmed that the ultrasonic waves accelerated the reaction better and improved the water treatment process as well as taste of water significantly. The experimental results also revealed that it is very difficult to compare the efficiency of water treatment process with the effect among irradiation times (30 min., 20 min. and 10 min.) quantitatively which warranting further examination on this.

REFERENCES

1. D. R. Pathak, A. Hiratsuka and Y. Yamashiki, "Influence of Anthropogenic Activities and Seasonal Variation on Groundwater Quality of Kathmandu Using Multivariate Statistical Analysis," In: Peters et al., Eds., Assessment of Water Quality under Changing Climate Conditions, Proceedings of Symposium H04 held during IUGG2011 in Melbourne, IAHS Publications, Vol. 348, No. 24, 2011, pp. 67-72.

2. Y. Magara, "Water Quality and Public Health in Japan after the Tsunami," Proceedings of Water Convention 2012, IWA, (2012), Session Theme 4: Water Quality & Health.

3. M. A. Sameraiy, "A Novel Water Pretreatment Approach for Turbidity Removal Using Date Seeds and Pollen Sheath," Journal of Water Resource and Protection, Vol. 4, No. 3, 2012, pp. 79-92.

4. U. I. Gaya and A. H. Abdullah, "Heterogeneous Photocatalytic Degradation of Organic Contaminants over Titanium Dioxide: A Review of Fundamentals, Progress and Problems," Journal of Photochemistry and Photobiology C" Photochemistry Reviews, Vol. 9, No. 1, 2008, pp. 1-12.

5. M. N. Chong, B. Jin, C. W. K. Chow, et al., "Recent Developments in Photocatalytic Water Treatment Technology: A Review," Water Research, Vol. 44, No. 10, 2010, pp. 2997-3027.

6. K. Fujita, “Shirarezaru Mizu no Chonouryoku (Unknown Supernatural Water Power) (in Japanese),” Kodansha, 2007, pp. 12-74.

7. Y. Saeki, “Orugouru ha Nou ni Kiku (Activation of Brain by Music Box) (in Japanese),” Jitsugyou no Nihonsha, 2007, pp. 92-120.

8. D. H. Oyib, “Ultrasound Technology in Water Treatment: Suppressing Algal Growth and Biofilm Formation,” 2013. http://www.water-world.com/articles/2009/05/ultrasoundtechnology-in-water-treatment-suppressing-algal-growthand-biofilm-formation.html

9. S. Hashimoto, “Evaluation of Water Quality of Healthy and/or Tasty Drinking Water and Its Application to Japanese Waters (in Japanese),” Journal of SHASE, Vol. 63, No. 6, 1989, pp. 463-468.

CHAPTER 2

Ultrasound generation with high power and coil only EMAT concepts

Dirk Rueter, Tino Morgenstern

HIGHLIGHTS

- Non-contact ultrasound from high power induction coils.
- MHz frequencies and relatively high lift off distances (up to 2 cm).
- High momentary pressures (up to 40 MPa or 400 atmospheres) onto various metals.
- Description of the physics by handy and integral expressions.

ABSTRACT

Electro-magnetic acoustic transducers (EMATs) are intended as non-contact and non-destructive ultrasound transducers for metallic material. The transmitted intensities from EMATS are modest, particularly at notable lift off distances. Some time ago a concept for a "coil only EMAT" was presented, without static magnetic field. In this contribution, such compact "coil only EMATs" with effective areas of 1–5 cm^2 were driven to excessive power levels at MHz frequencies, using pulsed power technologies. RF induction currents of 10 kA and tens of Megawatts are applied. With increasing power the electroacoustic conversion efficiency also increases. The total effect is of second order or quadratic, therefore non-linear and progressive, and yields strong ultrasound signals up to kW/cm^2 at MHz frequencies in the metal. Even at considerable lift off distances (cm) the ultrasound can be readily detected. Test materials are aluminum, ferromagnetic steel and stainless steel (non-ferromagnetic). Thereby, most metal types are represented. The technique is compared experimentally with other non-contact methods: laser pulse induced ultrasound and spark induced ultrasound, both damaging to the test object's surface. At small lift off distances, the intensity from this EMAT concept clearly outperforms the laser pulses or heavy spark impacts.

KEYWORDS

EMAT; Pulsed power; Non-linear effects; Spark induced ultrasound; Laser induced ultrasound

1. INTRODUCTION

Conventional electromagnetic acoustical transducers (EMATs) include a permanent magnet and an induction coil. The coil induces RF eddy currents into a metallic test object. The RF eddy currents and the presence of the permanent field B_0 result in oscillating Lorentz forces, i.e., ultrasound excitation in the metal. The method excites metallic test objects over some distance via immaterial magnetic fields. As an appreciated feature, this magnetic ultrasound transduction readily permeates most non-metallic barriers (air, oxides, oil, painting, humidity, packaging material: paper, plastic foil, etc.). A variety of possible geometries and ultrasound modes (shear waves, longitudinal waves, etc.) can be applied. As a fundamental problem, the transmitted ultrasound intensities from EMATs are modest. Usually, only very small "lift off" distances in the range of less than few mm are practical for proper EMAT operation. On the other hand, many application fields – like heavy steel industries or prepackaged metal parts – would appreciate increased lift off distances or higher ultrasound intensities for crude and large test objects. Additionally, permanent magnetic fields from conventional EMATs attract ferromagnetic particles. Such adhering particles potentially disturb the measurement or even may cause mechanical damage to the transducer or the test object. Some recent EMAT types utilize pulsed electromagnets for B_0, thereby avoiding that problem (Refs.[1], [2], [3] and [4]).

In this contribution "coil only EMATs" are investigated, i.e., a permanent magnetic field is lacking. Instead, the RF induction field itself interacts with excited (and orthogonal) eddy currents in the test object, producing Lorentz forces and acoustic pressure in the sample. This concept was already investigated by Jian et al. (Refs. [5] and [6]). The authors modeled and experimentally verified (Ref. [5]) the ultrasound emission from two different EMAT systems (spiral coil and line coil). Lorentz forces and displacements in the test metal resulted from eddy current and a permanent (static) field B_0 and additionally – of particular interest here – from eddy current and the RF induction field itself (i.e., the dynamic field). With 0.1 mm lift-off, a static field B_0 = 0.395 T and a characteristic excitation frequency around 300 kHz the authors found (cited literally from Ref. [5]) "the force due to the dynamic field is about five times larger than the Lorentz force due to the applied static magnetic of 0.395 T". This remark is very interesting, since the permanent field B_0 in conventional EMATs cannot be easily increased by a factor of five: for Ref. [5]this would be a quite challenging 2 T and that is currently not possible with conventional magnet materials.

EMAT schemes without a static field were also mentioned in even older textbooks (e.g., Ref. [7]). Herein, the occurring Lorentz forces are recognized to be exclusively repulsive, they increase with the square of excitation current and they oscillate with a doubled frequency. For relatively low frequencies (100 kHz…) considerable ultrasound intensities (at doubled frequency) were quoted, although not explicitly quantified. The lacking permanent magnet allows more compact and robust designs (e.g., no critical temperature limits of certain magnetic materials). As a clear disadvantage (Ref.[6] and [7]), without a static field this EMAT scheme cannot readily detect ultrasound, it then exclusively works as a transmitter. Possible ways to overcome this inherent obstacle are discussed below in the conclusions.

Additionally, a good understanding and much experience about strong Lorentz forces and high pressures from "coil only transducers" already has existed for decades in the well-established field of electromagnetic forming (EMF) (Ref. [8]). In EMF not only pressures onto a metal are generated over a distance ("non-contact") via pulsed magnetic fields. Additionally, these forces are strong enough to overcome the structural strength of the work piece. Then the metal gives way to the pressure pulse and undergoes a rapid and considerable deformation. EMF is industrially exploited as a non-contact tool for joining and forming metallic work pieces (Refs. [9], [10] and [11]). The characteristic frequency range of a pulse for EMF is in the order of just 10 kHz, corresponding to several tens of μs for a single half wave. These kHz frequencies are chosen to optimize the overall EMF process and its efficiency. The generation of ultrasound is not intended in EMF, although recognized as a by-product. The very strong Lorentz forces in EMF – readily beyond limits for non-destructive testing (NDT) – encourage the consideration of this principle for intense and non-contact ultrasound generation.

Our intention is the investigation of powerful and magnetically induced (non-contact) ultrasound at notably higher frequencies (i.e., >1 MHz instead of 300 kHz or less), suitable for NDT. Furthermore, increased lift off distances towards 1 cm instead of just 1 mm or less are approached, interesting for more or new application fields like prepackaged products. The underlying physics is viewed under the terminology*magnetic pressure* (more common in EMF technologies) instead of explicitly evaluating eddy currents and Lorentz forces (as typically done in EMAT considerations). The approach via magnetic pressure appears to be more convenient here and it still reasonably describes the actual findings from experiments.

The well-known fundamentals of EMF also apply at MHz frequencies and can be briefly explained by Fig. 1, Fig. 2 and Fig. 3 and some robust relations. For EMF technology, such relations and approximations are found to sufficiently match the experimental observations and in addition, they are quite lucid and handy (Ref. [11]). Due to the complex geometries of magnetic fields and eddy currents in EMAT geometries, an analytically exact description can just be derived for simple geometries (such as a current line or a current loop parallel over a metallic plane). Due to a more extensive and abstract evolution this way is not approached in this practically motivated contribution. Instead, the problem is modeled in a FEM computation (as also carried out in Ref. [5]). The FEM modeling yields a numerical correction for some quite basal approximations and finally, with this correction, the experimental values are found in reasonable accordance:

A capacitor C is charged to high voltage levels U. The stored energy E in the capacitor is (Eq. (1a))

$$E = \frac{1}{2} C \cdot U^2 \tag{1a}$$

and

$$E = \frac{1}{2} L \cdot I_{el}^2 \tag{1b}$$

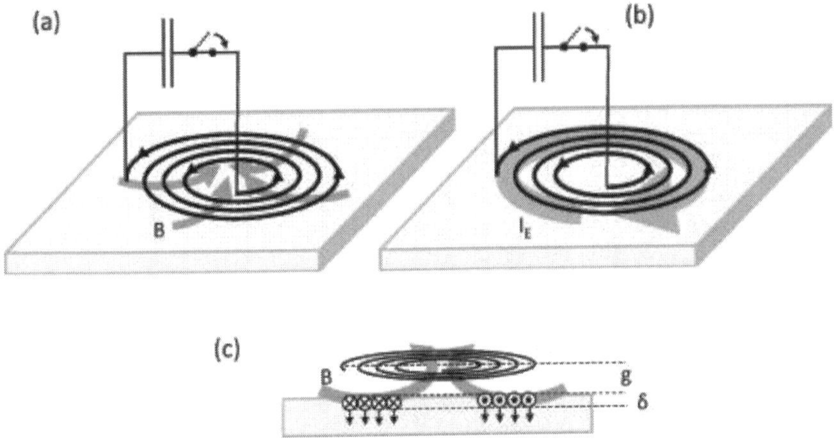

Figure. 1. A capacitor discharges into a flat spiral coil. The coil is positioned close to a metallic test object. The released energy converts into a magnetic field *B*, this field is directed radially and parallel to the metallic surface (a). A contra-directional eddy current I_E is induced in the metal beneath the coil (b). The magnetic field B and the eddy current are restricted to the skin depth δ and result into Lorentz forces or a pressure normally into the depth of the metal (c). The effect works over some distance *g*.

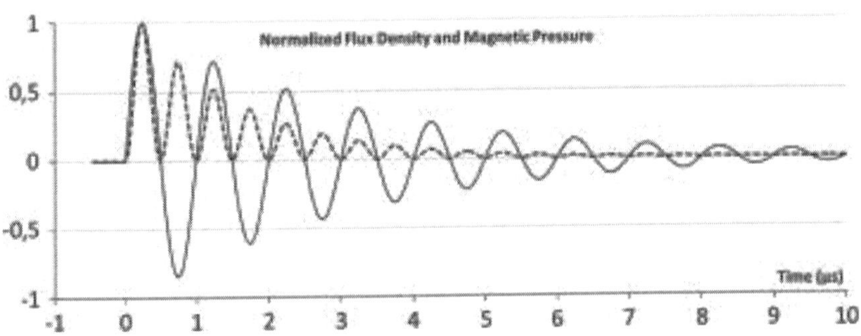

Figure. 2. At 0 µs a damped 1 MHz oscillation (solid line) of the magnetic flux density *B* is started. The energy density or magnetic pressure is proportional to B^2 and is displayed by the interrupted line. The pressure exclusively obtains positive polarity and it oscillates at 2 MHz. Idealized and normalized representation from a calculation, $\tau = 3$ µs.

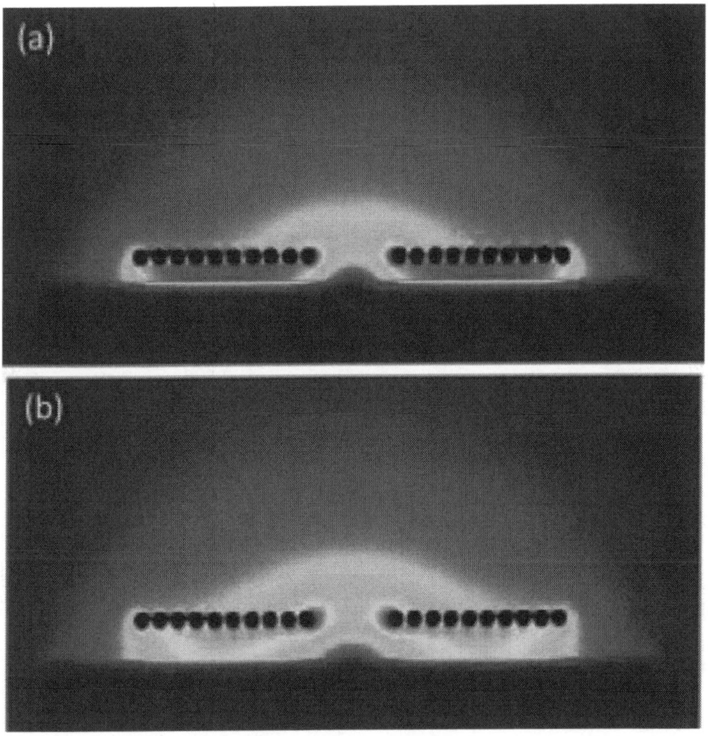

Figure. 3. Magnetic field intensity in intuitive colors from a flat spiral coil at $g = 1$ mm (Fig. 3a) and at $f = 1$ MHz, over a highly conductive copper block (lower rectangle). Cross sectional view similar to Fig. 1c. The B field virtually cannot permeate into the metal but is shielded (or compensated) by eddy currents. Therefore the field is concentrated in the gap g, the field direction is radial and parallel to the metal surface (not visualized here). In Fig 3b the gap g is doubled to 2 mm and the current in the coil is unchanged. Now the field intensity in the gap is notably decreased and the field tends to localize around the coil. Images generated from simulation.

In commercially available EMF machines E reaches or even considerably exceeds 10 kJ. After closing a switch (Fig. 1), the capacitor C will rapidly (in EMF: within 10^{-5}–10^{-4} s) discharge into an inductor L. Such an inductor L can be realized as a flat spiral coil. For ideal elements (switch, L and C) the stored energy E is fully converted into a magnetic field of the inductor L with discharge current I_{el}, described by Eq. (1b). The arrangement represents a simple L-C-oscillation circuit; the current in the conductor (and the associated magnetic field) then will oscillate with a characteristic frequency f. The characteristic electrical impedance Z_{el} is the ratio of voltage amplitude and current amplitude in the oscillating L-C-circuit:

$$f = \frac{1}{2\pi\sqrt{L \cdot C}} \tag{2a}$$

$$Z_{el} = \sqrt{\frac{L}{C}} \tag{2b}$$

For NDT ultrasound, L-C-frequencies in the order of 10^6 Hz are interesting and this can be readily achieved by choosing elements L and C with relatively smaller values. The available maximum electrical power $P_{el,max}$ from such oscillating L-C-circuit is

$$P_{el,max} = \frac{1}{4} \cdot \frac{U^2}{Z_{el}} = \frac{1}{4} \cdot \frac{1}{\sqrt{L \cdot C}} \cdot C \cdot U^2 = \pi \cdot f \cdot E \tag{3}$$

When extracting this available maximum power (Eq. (3)) from the L-C-circuit, the oscillating energy (Eq. 1) will essentially be consumed in less than a single half-wave.$P_{el,max}$ is used below as a reference power for efficiency considerations.

In practical setups, neither the coil nor the switch nor the capacitor is ideal. The energy losses and resistance of these elements result into a damped oscillation of the current or the magnetic flux density B

$$B(t) \sim \sin(2\pi ft) \cdot e^{-\frac{t}{\tau}} \tag{4}$$

where $\tau = L/R$ and R is an equivalent serial resistance in the L-C-circuit to represent energy losses after closing the switch. Fig. 2 displays a damped 1 MHz oscillation with$\tau = 3$ µs.

When positioning a flat spiral coil L close to a metallic object in a distance g (the lift off gap), the magnetic field B has a mainly radial component parallel to the metal surface. The oscillating field will induce an annular and contra-directional eddy current in the metal below the spiral (Fig. 1a and b). This eddy current couples back to the spiral coil and affects the frequency f (usually increases) and the decay time τ (usually decreases) of the L-C-circuit. Particularly at very close coupling between inductor and metal noticeable changes result. The general L-C-behavior however, is not principally affected at moderate coupling, and for the sake of clarity these effects are not explored in more detail here, see instead a focused elaboration in Ref. [6].

It is well known that an RF eddy current is carried by just a thin surface layer of the metal, with effective thickness δ ("skin", skin effect) (see e.g. Ref. [12]). At 1 MHz, the characteristic thickness δ is about 85 μm for aluminum, about 400 μm for (non-ferromagnetic) stainless steel and for ferromagnetic steel about 10 μm (if magnetic flux density below saturation, up to 1.5 T) or about 200 μm (flux density above 2 T) (Ref. [12]). Neither the eddy current nor the RF magnetic field can permeate the metal much deeper than the skin depth δ: the induced and contra-directional eddy current itself generates a magnetic field. That additional field finally eliminates (subtracts) all resulting magnetic field below the skin depth. It increases (adds, direction reversed) the resulting magnetic field *above* the surface, within the lift off gap. Furthermore, *above* the excitation coil (here the excitation field is reversed) the eddy current again weakens the field. Therefore, the resulting magnetic field B at RF frequencies and in an arrangement according to Fig. 1 is practically forced above and parallel to a metal surface, and in addition, the highest flux densities are present in the lift off gap g.

It should be noted that for a given RF magnetic field the total eddy current is – even for different metal types with different conductivity – virtually the same. This behavior implicitly results from the compensation – by eddy currents – of all induction fields in the volume of a metal. Fig. 3 illustrates a FEM simulation for a magnetic flux density. The computed scenario displays the cross section of a "flat spiral coil with RF excitation above a metallic surface". The intensity of B is intuitively represented by the color. Near the metal, the B field is directed parallel to the surface, the direction itself is not visible in this presentation. The B field does virtually not permeate the metal; it is restricted to the relatively thin surface layer δ. In the gap g between metal and coil the field appears relatively strong and (particularly for Fig. 3a) fairly homogeneous. Above the coil the field is much weaker; the field is not symmetric with respect to the coil plane. For different kinds of metal (different skin depth δ) the general appearance of Fig. 3 does virtually not change, as long as the gap g between coil and metal is significantly higher than skin depth δ (which is implied in this simulation).

As a further remark, the virtually equal eddy currents for quite different types of metals (good conductivity vs. poor conductivity) do not imply an equal energy consumption from the excitation coil: The phase between voltage and current of the inductor system is shifted below 90° by imperfect conductors (either caused by the imperfect coil itself or by the inductively coupled target metal), thus increasing the effective serial resistance R and reducing the τ (Eq. (4)). Actually, a reduced τ *for* a target metal with relatively poor conductivity – here stainless steel – is experimentally observed below.

When comparing the modeling of Fig. 3a and b (same excitation current implied), the field in the smaller gap of Fig. 3a appears to be more concentrated and more homogeneous. In addition, the residual field *above* the coil plane is weaker than in Fig. 3b. The field intensity around the coil is considerably deformed by the superimposing field from the eddy current. When numerically integrating the magnetic energy density (being proportional to B^2, see from Eq. (5) below) over the volume in the FEM models, for Fig. 3a about 69% of the total magnetic field energy is accumulated in the volume below the coil center plane. Actually it is concentrated between the coil windings and the metal surface. For the scenario with the increased gap (Fig. 3b) the relative energy content in the gap accounts to a somewhat less 65%, relatively more field energy is then also present above the coil, apart from the target.

As a reasonable approximation – derived from such FEM data (Fig. 3a and b) and as usually done in EMF techniques – it can be stated, that in a scenario where the gap g is smaller than 1/10 of the active radius of a flat spiral coil, the magnetic energy of the system (Eq. (1b)) will predominantly (i.e., >70%) be present in the gap.

The annular eddy current and the mostly radial magnetic field – both parallel to the metal surface – result into Lorentz forces. These distributed forces are directed normally into the depth (Fig. 1c) and are experienced as a pressure *p*. As an important detail, this pressure is also positive for a negative half wave of the oscillating magnetic field (Fig. 2and Refs. [5] and [7]). Not only the *B* field but also the accompanying eddy current periodically changes the polarity, only positive pressure is produced (as intended in EMF). As an inherent consequence the Lorentz forces or pressure then will then oscillate at a doubled frequency (Ref. [7]). Furthermore, the Lorentz force is proportional to the magnetic flux density *B* and the eddy current, which itself is proportional to *B*. Then the Lorentz force and the experienced pressure is proportional to B^2 (Ref. [7]).

It is well known in EMF and other disciplines like magneto-hydrodynamics (Ref. [13]) that this experienced pressure *p* actually equals the *energy density* of an oscillating or transient magnetic field parallel and in contact to a metallic surface:

$$p = \frac{1}{2} \cdot \frac{B^2}{\mu_0} \qquad (5)$$

$_0$ represents the magnetic permeability of the vacuum. The energy density of a magnetic field is, therefore, also denoted as the *magnetic pressure p*. It readily manifests as an ordinary mechanical pressure (Ref. [13]), when exposing sufficiently conductive matter (i.e., the skin depth δ is small with respect to other geometries, compare Fig. 3) to a transient magnetic field. Interesting for acoustics, with just the known magnetic pressure *p* or the known magnetic flux density *B* in vicinity to a good conductor (metal with relatively small skin depth), a more detailed and complex determination of eddy currents and finally resulting Lorentz forces is not necessarily required. And, since most of the pressure is directed normally into the depth of the metal, a preferred excitation of longitudinal waves, propagating normally into the depth, can be expected. This is already described by previous workers, here cited from Ref.[5]: “the out-of-plane Lorentz force (into the depth) due to the dynamic field is about 10 times larger than the (lateral or radial) in-plane force.” Only relatively small B field components are directed normally into the depth of the metal, basically located at the inner and the outer radius of the spiral coil. Here the field penetrates and then leaves the thin skin layer, and together with the annular eddy current some – relatively small – Lorentz forces in radial direction are generated.

At B = 1 Tesla the magnetic pressure is close to 400 kPa or 4 atmospheres (Eq. (5)) and this equals an energy density of 0.4 J per cm^3. In industrial EMF, an energy amount of about 15 kJ is converted into magnetic field energy. The magnetic energy, being proportional to B^2, is mainly (>70%) located and concentrated in the gap between coil and metal surface. Given that 10 kJ is homogeneously distributed in a volume of 100 cm^2(here just taken as an example for the effective free space between coil and work piece in EMF applications), the mechanical pressure on the metal then would be 1000 atmospheres. This can rapidly deform a metallic work piece. That pressure is equivalent (Eq. (5)) to a magnetic flux density B = 16 T. In fact, even higher pressures are reported in EMF techniques.

It should be noted that the limiting problems in the EMF technique arise from the vast Lorentz forces acting on the coil windings rather than just heat due to the very strong current pulses. Furthermore, all known ferromagnetic materials are saturated above$B$ = 2 T; thereby an iron or ferrite core will not significantly improve these techniques. The work is usually done with compact and rigid “copper alone” coils. As an interesting side remark, ferromagnetic iron will not be attracted by a strong magnetic induction field but it will be repelled, as all other non-magnetic metals.

The effective sound intensity I of an ultrasound wave at sound pressure p (peak value) in a material is

$$I = \frac{p^2}{2 \cdot Z_M} \tag{6}$$

with the characteristic acoustic impedance Z_M. At first sight, when implying a direct conversion of oscillating magnetic pressure into sound pressure, the excited sound intensity I should increase with B^4 or the stored energy E^2 (just by combining Eqs.(5) and (6)). This is a quadratic effect or an effect of second order. The relation directs the way towards non-contact and high intensity ultrasound generation via a "coil only transmitter": the ultrasound power should increase with the square of excitation power.

As an example, the magnetic pressure and the acoustic pressure at B = 1 T are given as 400 kPa (0.4 J/cm^3) and 200 kPa respectively. Only 200 kPa can be utilized for the acoustic pressure, because the full wave (peak to peak, from –200 to +200 kPa) must be within the magnetic amplitude of 400 kPa (see Fig. 2). The sound intensity I of a longitudinal bulk wave in aluminum ($Z_M \approx 17 \cdot 10^6$ Ns/m^3) then would be 1175 W/m^2 (Eq.(6)) or roughly 0.12 W/cm^2. At a given volume of 0.1 cm^3 for the magnetic energy (from spiral coil at 1 mm lift off gap g and 1 cm^2 footprint area A) and a frequency of 1 MHz the available power $P_{el,max}$ in the LC – circuit should be at least 40 mJ $\cdot \pi \cdot$ 1 MHz = 125 kW (from Eq. (3)). This accounts for the ideal case that all magnetic energy is exclusively and homogeneously allocated within the gap. Actually, only about 70% of the energy is in the gap (see the FEM discussion above) and secondly, the field is not exactly homogeneous in the gap but gradually weakened towards the metal target. This local inhomogeneity will, evaluating the FEM data in the gap and the B^2 dependence of the pressure, result in an additional 70% drop. Together these real effects account to a pressure reduction of approximately 50%. Then for 0.12 W ultrasound at 2 MHz in aluminum an available power of about 250 kW at 1 MHz should be present in the LC-circuit. Now a flux density of 1 T can be expected onto the metal. This suggests poor conversion efficiency and a vast mismatch situation; the vibrating metal surface can only convert about 0.5 ppm from the available power.

When applying a 100-fold higher electrical power density in the gap, i.e., an available power of 25 Megawatts at 1 MHz, the energy and the magnetic or acoustic pressure also increases 100-fold. As a result (Eq. (6)) a much more powerful ultrasound signal – now about 1.2 kW – is expected, corresponding to a magnetic flux of 10 T near the metal and a peak pressure of 40 MPa. Furthermore, the efficiency is improved significantly to 50 ppm, but is still not very good. To the benefit of this number, that available maximum power as a reference level is not identical to the actual power dissipation (which is considerably smaller) of the total system, and the local power dissipation in the metal is even much smaller.

It should be noted that this particular example (25 MW available power at 1 MHz,g = 1 mm and A = 1 cm^2, resulting into 40 MPa peak pressure) is – within reasonable accuracy – practically verified in the experiments below.

In a general formulation of the above considerations, the finally transmitted ultrasound power P_{US} of a ringing LC-circuit with released energy E, lift off g and coil area A over a metal with acoustic impedance Z_M can be stated as

$$P_{US} = \left(\frac{E}{g \cdot A} \cdot \frac{1}{2} \cdot \frac{1}{2}\right)^2 \cdot \left(\frac{1}{2 \cdot Z_M}\right) \cdot A \tag{7}$$

This formula should be fairly accurate for a situation, where g is about 1/10 of the coil's diameter (or the square root of the footprint area A), see FEM results in Fig. 3 and considerations above. And, on the other hand, the skin depth δ should be notably smaller than g. Inside the left bracket is the magnetic pressure, equivalent to the magnetic energy density: total energy E over active volume and then reduced to 50% by real effects, derived from FEM analysis. Another 50% reduction of the magnetic pressure is caused by the just 50% utilization of the only positive amplitude (Fig. 2) for bipolar acoustic amplitudes. Then the resulting acoustic pressure in the left bracket is squared and divided by 2 Z_M (the second bracket) and this is just the sound intensity according to Eq.(6). Actually, with a released energy of 8 J at $f = 1$ MHz, equal to an available power of 25 MW (Eq. (3)), and the geometries of $g = 1 \cdot 10^{-3}$ m and $A = 1 \cdot 10^{-4}$ m^2, in aluminum a momentary ultrasound power of 1175 W (equal to a momentary peak pressure of 40 MPa over 1 cm^2) is obtained from Eq. (7) and this was already discussed in the example above.

An efficiency η as the ratio P*US*/P*el,max* can be readily derived from Eqs. (7) and (3):

$$\eta = \frac{P_{US}}{P_{el,max}} = \frac{1}{32} \cdot \frac{E}{Z_M \cdot A \cdot g^2 \cdot \pi \cdot f} \tag{8a}$$

Eq. (8) can be altered by replacing the corrected energy density $E/(2 \cdot A \cdot g)$ by an energy density $B^2/2\mu_0$:

$$\eta = \frac{1}{32} \cdot \frac{B^2}{Z_M \cdot \mu_0 \cdot g \cdot \pi \cdot f} \tag{8b}$$

Eq. (8b) describes the efficiency as a function of the flux density B in vicinity to the metal. There are similarities to an expression reported much earlier by Dobbs (Ref.[14] and [15]):

$$\eta = \frac{P_{US}}{P_{RF}} = \frac{B_0^2}{Z_M \cdot \mu_0 \cdot \delta \cdot \pi \cdot f} \tag{9}$$

Despite the obvious similarity, the equation from Dobbs was derived in a quite different way, it has a somewhat different meaning and it applies for conventional EMATs with static field B_0.

When directly comparing Eqs. (8b) and (9), there appears a factor of 32 to the benefit of conventional EMATs. Another substantial difference to Eq. (8b) is that the skin depth δappears instead of the lift off gap g. Since $\delta \ll g$ in relevant applications and together with the factor of 32, the efficiency of a conventional EMAT (Eq. (9)) seems to be very much higher than in the coil only concept from above (Eq. (8b)). This is however not the case, as briefly discussed now.

It must be noticed that in Eq. (9) just the *absorbed* power P_{RF} (Refs. [14] and [15]) is taken as the reference level for the ultrasound power P_{US}. Eq. (9), therefore, only addresses the *local conversion efficiency* within the test metal, with a somehow given field B_0 and a somehow inductively inserted net power P_{RF}. In contrast, Eqs. (8b) and (8) offers a global description for the total setup, including the distant "coil only EMAT" and real attenuation effects over a distance g. And additionally, a much higher *available maximum* power in the LC-oscillation circuit is taken as the reference level for the conversion efficiency. The available maximum power in an oscillating LC-circuit even is considerably higher than the actual power dissipation (depending on τ, Eq. (4)) in the total system.

Particularly and to the benefit of coil only EMATs, the RF *B*-field cannot permeate into the depth of the metal; the field energy is concentrated close to and above the surface (Fig. 3), and this effect notably increases the energy density – equivalent to pressure. To the contrary, a static field B_0 from conventional EMAT will expand into the metal and this considerably dilutes the energy density B_0^2. For partial compensation the field source B_0usually must be oversized, both in geometry (resulting in less compact designs) and total magnetic field energy. Therefore, the skin effect for an RF B-field should not be seen as a disadvantage but as an advantage. The skin effect helps to concentrate the B field in the region of interest. In particular, very high flux densities above 2 T (hardly possible for conventional EMATs) can be realized in the relatively smaller volume, see EMF and see the observation in Ref. [5] with the superior dynamic field. Additionally, the regions of high induction field automatically coincidence with the regions of high eddy currents. This is not necessarily the case for conventional EMATs with separated field generation.

In another substantial difference to normal EMATs the coil only concept will – since only positive pressures result (Fig. 2) – transfer a linear momentum (“LM”) to the metallic test object. The test object will be repelled by the inductor and this is the intentional purpose in EMF technologies. A linear momentum can be easily determined by measuring the (constant) velocity of a mass, after momentum transfer by the coil. The linear momentum LM is delivered within a few periods of the MHz oscillation, Fig. 2 suggests the completion of the process after a few μs.

Since in Fig. 2 the maximum pressure p (and also the maximum force F) is normalized to “1 N/m^2” or “1 N” for the damped oscillation with $f = 1$ MHz and $\tau = 3$ μs, the normalized linear momentum (“NLM”) from this magnetic pressure oscillation is

$$NLM = \int_0^\infty F(t)dt = \int_0^\infty 1N \cdot \sin^2(2\pi f t) \cdot e^{\frac{-2(t-\frac{1}{4f})}{\tau}} dt \approx 0.88 \cdot 10^{-6}\,\text{kg}\,\frac{\text{m}}{\text{s}} \tag{10}$$

For an experimental high power oscillation with frequency f and decay time τ – which both are easily observable with a standard oscilloscope – the real linear momentum LM of the repelled test object can be set in relation to the normalized momentum NLM. The ratio of real LM to NLM equals the ratio of real maximum force F_{max} to 1 N. For 1 MHz and$\tau = 3$ μs:

$$\frac{F_{max}}{1N} = \frac{LM}{0.88 \cdot 10^{-6}\,\text{kg}\,\frac{\text{m}}{\text{s}}} \tag{11a}$$

In more general, for any other frequencies f and other decay times τ of a pulsed LC-discharge:

$$F_{max} = \frac{LM}{\int_0^\infty \sin^2(2\pi f t) \cdot e^{\frac{-2(t-\frac{1}{4f})}{\tau}} dt} \tag{11b}$$

Quantitative information about F_{max} is very welcome, since hereby also the practical and maximum magnetic pressure p_{max} is reasonably accessible via the affected area A. Obviously, p_{max} is tied to the achieved sound pressure in the metal.

p_{max} can be compared with the total energy E in the system (stored in the capacitor Cbefore switching), in best case it should equal the energy density in the gap g with area A. In practical setups and as discussed above, with a distance g $\approx$ 1/10 ·square root (A), the energy density close to the metal is expected to be only 50% of the best case limit $E/g \cdot A$.

2. TECHNIQUES, EQUIPMENT AND EXPERIMENTAL SETUP

50 high voltage (40 kV max) and 3 nF ceramic pulse capacitors (Murata DHS N4700) were combined in parallel to 150 nF. Such ceramic capacitors are commonly used e.g. in pulsed excimer lasers (Ref. [16]), where the stored energy must be delivered within some tens of nanoseconds and, consequently, at power levels of many Megawatts and currents in the kA range. These pulse capacitors obtain low internal inductance and resistance and they are proven to deliver many millions of pulses over their lifetime without fatigue or failure.

The multiple capacitors were interconnected with a low inductance wiring (short and wide strips). At an experimental charge voltage of 15 kV the stored energy E is close to 17 J (Eq. (1)). A standard high voltage DC supply with an adjustable current up to 20 mA and adjustable voltage up to 30 kV was used as a charge supply for the 150 nF. It is connected over a high voltage resistor (500 Ω) for decoupling the pulsed power from the DC supply.

For a desired oscillation frequency of 1 MHz, a total inductance L_T = 170 nH (including the parasitic inductance L_P from all wiring, switch, capacitor) is required (Eq. (2)). The spiral coil L_I, therefore, must be less than 170 nH. Since the stored energy E in capacitor bank C is converted to a major portion into magnetic field energy $E = L/2 \cdot I^2$ (Eq. (1b)) the maximum discharge current after closing the switch should then be close to 14 kA. The characteristic electrical impedance Z_{el} of such L-C-oscillation circuit is then 15 kV/14 kA ≈ 1 Ω (or alternatively by Eq. (2)). All real resistance in the LC-circuit must be held significantly below 1 Ω for an oscillation without excessive attenuation: Eq. (4),$\tau = L/R > 3$ μs, better: $\tau > 4$ μs. The available maximum power (Eq. (3)) is then close to 52 MW at stored energy $E = 17$ J.

Not all power or magnetic energy is present in the spiral coil L_I, because the energy is proportionally divided to the all serial inductances $L_T = L_I + L_P$ in the system. Without spiral coil L_I (shortcut) and from parasitic L_P only, the observed oscillation (or resonance) frequency of the system is 1.43 MHz, and (Eq. (2)) $L_P \approx$ 83 nH. In consequence, L_I must be 85–90 nH to meet the 1 MHz and only about 50% of the total energy or available power is present in the flat spiral coil, about 8 J. Then it can be reasonably estimated that only about 25 MW or less can be provided from the spiral coil. Without load (not coupled to metal target), a $\tau \approx 4$ μs is observable. Therefore all resistances in the system (switch closed) add up to less than 0.1 Ω.

The high power oscillation event can be readily observed from a distance of about 20 cm with a single loop of copper wire. The diameter is about 5 cm and the loop is loaded with a very low inductance 0.2 Ω resistor (i.e., a bundle of many resistors in parallel). The voltage over the 0.2 Ω is proportional to the current in the loop and can be readily observed via standard coax cable with a standard oscilloscope. Actually, the current in the loop is predominantly an eddy current from the distant LC-pulse generator, and it resembles the discharge current in the LC-circuit itself. With the observed oscillation and attenuation, a reasonable estimation of f and τ is possible. Due to the high momentary power no extra amplifier is needed, the signal from the loop (several Volts) is well suited for a standard oscilloscope. It must be noted that definitely an inductive signal is detected and not any other EMI effect: the magnetic coupling of the loop can be readily verified by just twisting the loop by 90° (plane then parallel to B). The signal then vanishes.

An important element for the pulsed LC-circuit (Fig. 1) is the switch. Required properties are low inductance ($\ll$150 nH), low resistivity ($\ll$1 Ω) and the capability to bear 15 kA and to withstand 15 kV. Furthermore, the switch must tolerate positive and negative currents during the oscillation and it must transit from high impedance ($\gg$1 M Ω) to low impedance ($\ll$1 Ω) within a time much smaller than 1 μs.

The requirements can be matched with commercially available thyratron tubes. Nowadays even an all-solid-state switching is established for similar applications, partly combined with magnetic pulse compression (compression in time). These all-solid-state techniques are maintenance free and have demonstrated a practically unlimited lifetime (Ref. [16]).

In our experiments, for the sake of simplicity and flexibility, a low inductance spark gap with sound absorber was applied, simply working in air. The switching time of such spark gap after ignition is much shorter than 1 μs (can be few nanoseconds) (Ref. [17]) and all requirements above (15 kV, 15 kA, $\ll$1 Ω after ignition) are well fulfilled. A certain amount of energy is consumed for the formation of very conductive and highly ionized plasma in the spark gap; this energy (estimated <5 J) is taken from the initial energy in the capacitor. For high power and "coil only" ultrasound generation a similar circuitry with a triggered spark gap was already presented in Ref. [7], although discussed only for smaller ultrasound frequencies (100 kHz...) and – together with a copper foil very close to the coil – intended as a mechanically contacting transducer for applications on (non-metallic) concrete. The basic disadvantage of such old-fashioned high power switch is the burn-off behavior after several hundred shots, requiring frequent maintenance. A silencer is mandatory: the plasma flashes emit considerable noise, naturally including ultrasound.

Therefore it is also interesting to compare the ultrasound pulses from a "coil only EMAT" with the impact of 17 J sparks itself from the capacitor bank. The sparks can be directly released onto the surface of the metallic test sample. Although representing a non-contact method over a certain distance (about 9 mm gap from electrode to surface), such spark technique cannot be called non-destructive. Considerable scorch marks affect the surface, and this cannot be tolerated in every application.

Additionally, a comparison with laser induced ultrasound is carried out (Refs. [18],[19] and [20]), from a pulsed Nd:YAG laser (Quantel Ultra 100) emitting at 1064 nm and 532 nm.

The laser pulse delivers 20 mJ energy within 8 ns and it is focused onto a small spot (<1 mm^2) of the test metal. As with the spark, the method is non-contact but however not "non-destructive": breakdown plasmas and impact marks affect the surface. These micro-explosions on the surface are apparently required for noteworthy ultrasound signals.

The flat spiral coils itself were made from standard PCB material with an extra thick copper layer (120 μm) or are, alternatively, manually wound spirals from regular copper wire. These spiral coils with a diameter of 25 mm down to 10 mm consist of just a few windings (<10) to meet the <170 nH requirement for f = 1 MHz. Due to the skin effect at 1 MHz in copper, the conducting material is not needed to be much thicker than 0.2 mm. In the pulsed power application the coils are exposed to high mechanical, thermal and electrical stress. The spiral coils were immersed in epoxy resin to mechanically stabilize and to electrically insulate the windings, since already two neighbored windings (=1 turn) must withstand a RF voltage difference of about 1000 V. Practically even more delicate is the voltage difference towards the test metal. Since an air gap (=the lift of distance g) of even almost 1 cm is overcome by the 15 kV, the surface area of the spiral coil is insulated by a 200 μm polymer foil (similar to Polyimide), laminated onto the coil with epoxy resin. With these insulating layers onto the surface, a gap g to a test metal of at least 0.5 mm is defined. An imperfection in the insulation – this already can be a small air inclusion – regularly results in instantaneous failure of the coil within one power pulse. On the other hand, a well handcrafted coil withstands a virtually unlimited number of shots.

Interestingly, the mechanical and thermal stability of the coils appears to be less delicate. The μs short and intensive mechanical forces can apparently be taken by the inertial mass of the high density copper material. Thermal problems become an issue at high repetition frequencies. At low repetition rates of about 1 Hz, a conventional air blower is sufficient for cooling a free standing coil. For higher repetition frequencies (10 Hz...) the back of the spiral coil was laminated onto a ceramic boron nitride block. This material is both a good electrical insulator and a good thermal conductor, thereby acting as a heat sink for the relatively small coil.

An aluminum rod with 30 mm diameter and 0.89 m length served as metallic test material, Fig. 4. The two end faces are plane. One face is exposed to a non-contact ultrasound generator, i. e., a coil only EMAT or the 17 J spark impact or the laser pulse. The other face was equipped with a ceramic PZT disk (10 mm diameter, 1 mm thickness), this piezo crystal was attached onto the aluminum face with paraffin wax. The two contacts of the piezo were loaded with 100 Ω and a parallel inductor of 3 μH for a prefered selection of 2 MHz and at 700 kHz bandwidth. This detector was directly connected to a standard oscilloscope; no extra signal amplifier was involved here.

An ultrasound shot into the rođs ending should result in a piezo signal at the other side after the characteristic runtime along 89 cm in aluminum: 0.89 m/6350 $ms^{-1} \approx 140$ μs. The delay time of 140 μs is well suited to distinguish strong electromagnetic interference (EMI) from a real ultrasound signal in the aluminum. The aluminum rod works as a waveguide and it actually disperses the ultrasound into several modes of longitudinal propagation. The investigation of this mode dispersion is however not within the scope of this contribution, see instead Ref. [21].

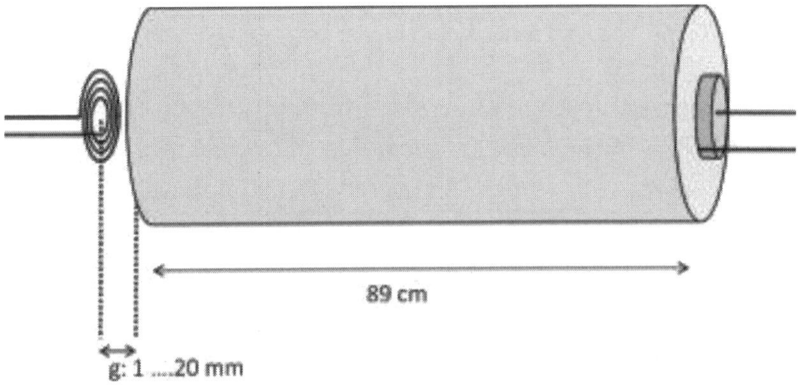

Figure. 4. Experimental setup. The spiral coil transmitter is positioned in a distance *g* to an aluminum rod. The other end of the rod is equipped with a piezo crystal and detects ultrasound events with a 140 μs delay from the 89 cm rod.

Furthermore, 30 × 30 mm platelets from aluminum, ferromagnetic steel with and without zinc coating and stainless steel (non-ferromagnetic) were prepared. These metal chips were placed, in a horizontal setup, onto the spiral coils. Within a magnetic pulse, a metal chip receives a linear momentum LM (Eqs. (10) and 15) and jumps against earth's gravity to a certain height *h*. This height h is readily observable with sufficient accuracy (only applies in square root of Eq. (16)), even with the naked eye. It must be noted, that an insulating material (e.g., a glass plate) does not jump. The LM is actually provided by the electromagnetic pressure and not by any mechanical vibration. The LM of the metal chip is

$$LM = m \cdot \sqrt{2 \cdot G \cdot h} \tag{12}$$

with m = mass of the metal chip, gravity $G = 9.81$ m/s^2, h = achieved height. With the known LM the maximum force F_{max} (Eq. (11b)) is derived, and then together with the inductive footprint area A, a characteristic magnetic pressure is accessible.

3. EXPERIMENTAL RESULTS

In Fig. 5 the oscilloscope screenshots from an 8 J pulse (i. e., a total 17 J energy in the system, including the parasitic inductances) through a spiral coil with approximately 1 cm^2 footprint A and 1 mm gap g to the aluminum rod are displayed. Channel 1 displays the piezo signal and channel 2 (also used for triggering) represents the induction signal in the wire loop.

In Fig. 5a clear piezo signal after 140 μs and with 15 V amplitude is present. In this moment, the piezo delivers more than 2 W of electrical power into the 100 Ω load. Note that the piezo covers only 10% of the rods face and it is not well coupled (just a paraffin wax layer with unintentional air inclusions) to the aluminum. Unquestionably the piezo detects a considerable ultrasound signal at the expected delay time. Even a multiple echo (distance then 280 μs, several times travelling forth and back over 180 cm) is observable, but not shown here. The discrete replicas in the ultrasound signal after 140 μs and at a characteristic distance of about 10 μs originate from different propagation modes (Ref. [21]) in the 30 mm rod, which also shall not be further discussed here.

Fig. 5b zooms into the electromagnetic excitation at t = 0 μs. A damped oscillation at 1 MHz is present in channel 1, the decay time τ resembles Fig. 2, and the amplitude halves every 2 μs. Therefore, τ can be reasonably estimated to be 3 μs here.

Fig. 5c zooms into the piezo signal after 140 μs. The predominant frequency is close to 2 MHz. The effective duration of a received ultrasound pulse signal (within a certain propagation mode) is 2–3 μs. The actual duration is assumed to be even shorter, since the piezo detector itself exhibits some time constant or limited bandwidth.

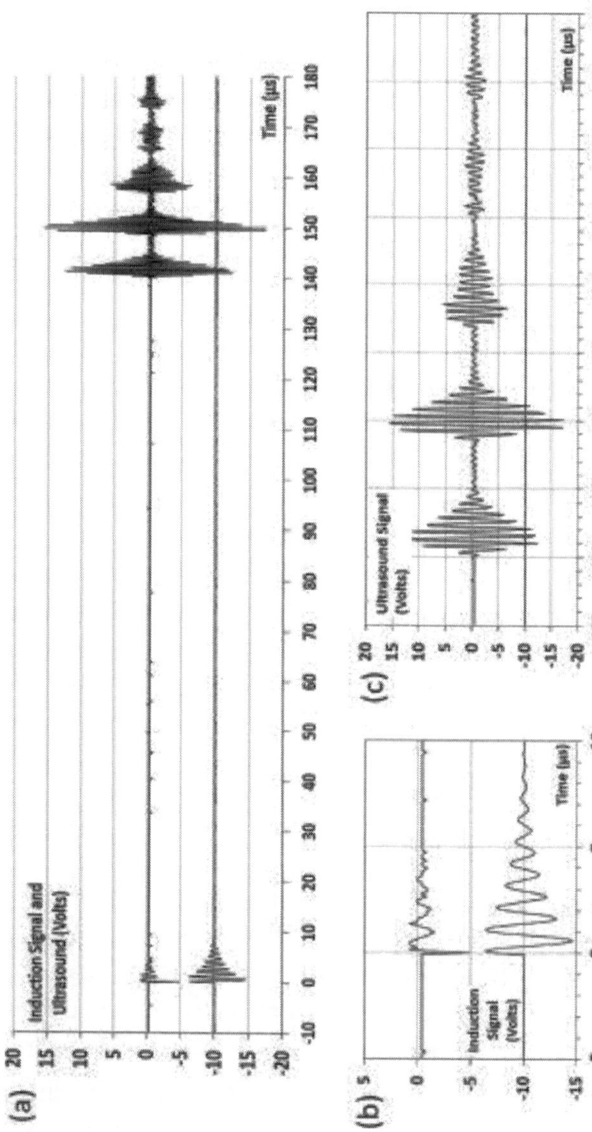

Figure. 5. Ultrasound signals (upper line) and induction signals (lower line) from a multi MW induction pulse at 1 MHz. Signal amplifiers are not required. At 0 µs the induction pulse and after 140 µs a strong ultrasound signal appears, separated in discrete propagation modes through the aluminum rod. In (b) the induction pulse. Besides an EMI affected first half wave, a regularly damped 1 MHz oscillation with $\tau \approx 3$ µs appears. The ultrasound signal in (c) obtains strong 2 MHz components.

When comparing the above ultrasound intensity (using the exact same piezo detector) with those signals from a direct 17 J spark impact into the aluminum rod or a nanoseconds laser impact, the electromagnetic method outperforms the other methods by far: the spark signal shows a very similar pattern in time after 140 μs (Fig. 5a) but achieves only 0.5 V amplitude and the laser pulse induced ultrasound (including plasma and resulting point defect on the aluminum surface) is just close to 0.03 V.

The electromagnetic induced ultrasound intensity *I* at 2 MHz and in W/cm^2 is, therefore, almost 1000 times stronger than from the direct spark impact with even doubled energy (17 J) and 250,000 times stronger than from the laser shot.

To better quantify the ultrasound intensities and pressures, 1 MHz electromagnetic "jump" experiments were made, here again with about 8 J energy in the transducer coil.Table 1 represents the data:

Experimental uncertainties exist with the experimental momentum LM (particularly the smallest jump from the stainless steel is not very accurate), with the characteristic decay time of the oscillation τ and with the actually affected area A. Nevertheless, a virtual independence of the experienced pressures from the metal type is obtained. The magnetic pressure implies a magnetic field of $B \approx 10.5$ T and a magnetic energy density of ≈4 J/0.1 cm^3 close to the metal. Actually, a total magnetic energy of about 8 J or an available power of 25 MW was present in the coil. The experimentally observed reduction of magnetic energy density down to only 50% close to the metal is in well accordance to the above considerations from the FEM model and the herein computed distribution of field energy. Furthermore, the jump data is – within the limited experimental accuracy of such delicate high voltage experiments – in quite reasonable accordance to the discussed example above, with the presumed 10 T and the equivalent 40 MPa.

Table 1. Electromagnetic jump experiments against earth's gravity. Different metal chips are vertically repelled to a certain height h by a spiral coil with $A \approx 1\ \text{cm}^2$ and $g \approx 1$ mm. The coil is excited with 2–3 µs short pulses at 1 MHz and 8 J energy, equivalent to 25 MW available power. Momentary pressures of about 40 MPa act on the metal chips, equivalent to a flux density of about 10 T.

Material and Size of the metal chip	Aluminum $30 \times 30 \times 1\ \text{mm}^3$	Ferromagnetic steel $30 \times 30 \times 0.55\ \text{mm}^3$	Stainless steel non-ferromagnetic $30 \times 30 \times 1\ \text{mm}^3$
Mass of the chip	$2.43 \cdot 10^{-3}$ kg	$3.91 \cdot 10^{-3}$ kg	$7.08 \cdot 10^{-3}$ kg
Jump height h	0.13 m	0.03 m	0.008 m
Experimentally observed Linear Momentum LM (Eq. (12))	$3.9 \cdot 10^{-3}$ Ns	$3.0 \cdot 10^{-3}$ Ns	$2.8 \cdot 10^{-3}$ Ns
Normalized linear momentum (NLM) (Eq. (10))	$\tau \approx 3$ µs $8.8 \cdot 10^{-7}$ Ns	$\tau \approx 2$ µs $6.4 \cdot 10^{-7}$ Ns	$\tau \approx 2$ µs $6.4 \cdot 10^{-7}$ Ns
Maximum force F_{max} (Eq. (11b))	4430 N	4700 N	4370 N
Maximum magnetic pressure p_{max} ($A \approx 1\ \text{cm}^2$)	$44 \cdot 10^6$ Pa ≈440 Atmospheres	$47 \cdot 10^6$ Pa ≈470 Atmospheres	$44 \cdot 10^6$ Pa ≈440 Atmospheres
Maximum acoustic pressure at 2 MHz	$22 \cdot 10^6$ Pa	$23 \cdot 10^6$ Pa	$22 \cdot 10^6$ Pa
Momentary acoustic intensity at 2 MHz (Eq. (6))	1.4 kW/cm²	0.6 kW/cm²	0.55 kW/cm²

Fig. 6 reveals the ultrasound amplitude at 2 MHz as a function of capacitor charge voltage (Eq. (1)) and at fixed geometry. The charge voltage ("kV") proportionally converts into flux density of the spiral coil ("**V** s/m^2") and the ultrasound power – as discussed in the introduction – should increase with B^4 and the sound pressure p with B^2. The piezo signal is proportional to the sound pressure or amplitude, and therefore, the piezo signal should increase with the square of the capacitor charge voltage. A parabola is expected and apparently, this is quite well matched. Also at 16 kV charge, a piezo signal of almost exactly 16 V is obtained (this value not included in the graph for better representation of the smaller values).

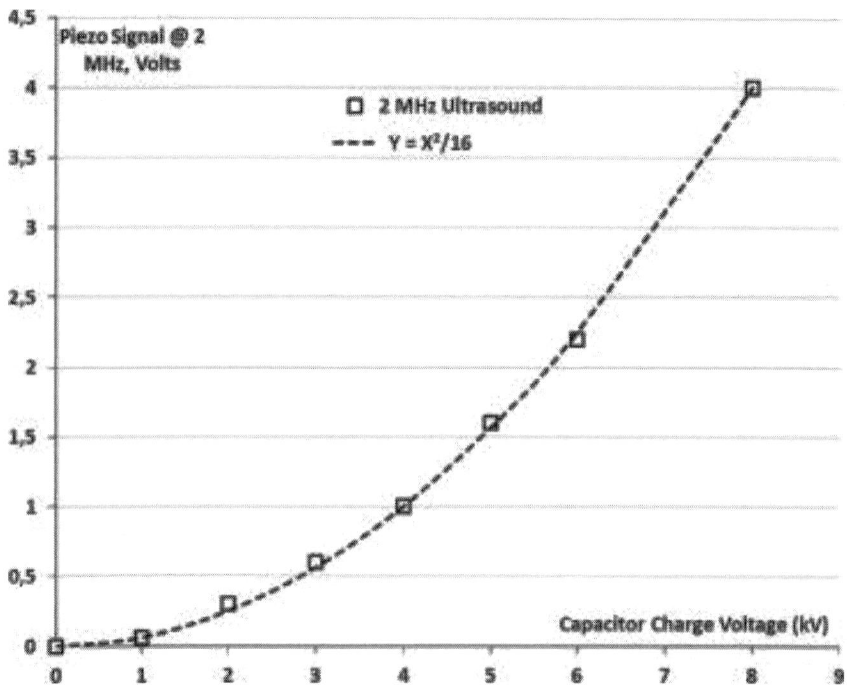

Figure. 6. Piezo amplitude at 2 MHz as a function of capacitor charge voltage. The observed data direct to a second order (quadratic) relation. Here for comparison a simple parabola $y = x^2/16$, dashed line. Actually, also at 16 kV charge a signal close to 16 V results. This is not shown here for better representation of the smaller values.

Fig. 7 shows the electromagnetically induced 2 MHz ultrasound after 140 µs as a function of "lift off distance" *g* for two different spiral coils. One coil has an effective diameter of 3 cm, the other 1.4 cm. The area *A* of the circles is respectively 7 cm^2 and 1.5 cm^2. These areas just can be counted for approximately 2/3, due to the inner bore in the spiral coil. The corrected *A* for the coils is then about 5 cm^2 and 1 cm^2.

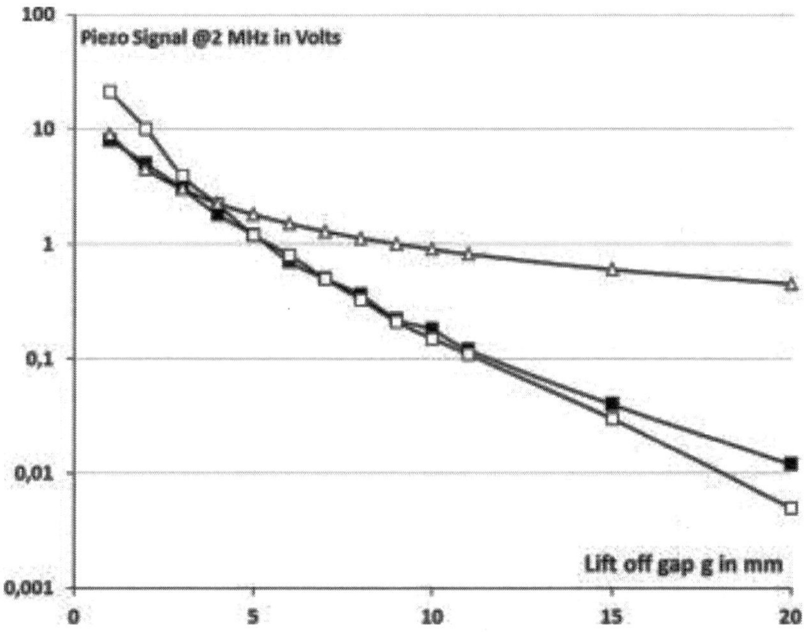

Figure. 7. Piezo amplitude at 2 MHz for a 1 cm^2 (white squares) and a 5 cm^2 coil (black squares) at constant energy and as a function of distance *g*. Towards increasing *g* the smaller coil is more affected. On the other hand, at small distances the smaller coil is more effective. For comparison, the functionality of Eq. (8) is displayed (Ultrasound Intensity ~ $1/g^2$, white triangles). At relatively low distances g the real behavior can be fairly matched.

At the same total energies, the ultrasound intensity and conversion efficiency is significantly higher for the smaller coil at $g = 1$ mm, following Eqs. (7) and (8) with the suggestion of the power density as a relevant parameter for efficient electroacoustic coupling.

With increasing gap g the ultrasound intensity rapidly decreases, almost exponentially. The smaller coil is affected more by the distance, so that above $g = 1$ cm the wider coil becomes superior. Actually the magnetic field energy is more localized around a smaller coil, all gradients are higher. In consequence, the magnetic pressure $\sim B^2$ of a small coil is more affected by distance than for a wider coil.

It must be noted that for higher distances g the B field becomes quite inhomogeneous in the gap, it gradually localizes towards the coil (Fig. 3b). Additionally the affected footprint area A will increase. Therefore a simple calculation of the behavior with Eq. (7) will not match for higher distances; the actual situation then becomes more disadvantageous. However at small distances (1–3 mm) and for the 5 cm^2 coil, the simplified relation from Eq. (7) seems to fairly match the experimental observations.

4. DISCUSSION AND CONCLUSION

The experimental findings, although not obtained in utmost accuracy due to the quite delicate handling with an open high voltage capacitor with considerable pulsed power capability, apparently support the key considerations of this contribution.

The jump experiment comes close to the example from the introduction, where from a 25 MW available power at 1 MHz a sound intensity I or power ($A = 1$ cm^2) of 1.25 kW at 2 MHz was predicted for aluminum, equivalent to 40 MPa peak pressure. In the experiments, the power offering was about 25 MW at 1 MHz and resulted to nominally 44 MPa peak pressure. The difference is – as we believe – reasonably small within the accuracy of all experimental uncertainties and approximations.

A noteworthy observation is the virtual constancy of maximum pressures p_{max}, independent from the metallic target material. This result, originating from similar eddy currents in even quite different metals, supports the general concept of magnetic pressure also for ultrasound. The more pronounced losses in steel or stainless steel due to higher skin resistance just result in more attenuation of the oscillation. Not the intensity but the pulse duration τ should become smaller. As an additional remark here, the nominal skin depth in stainless steel (≈400 μm at 1 MHz) is not very small with respect to the applied gap of 1 mm. But nevertheless, the observed momentum in stainless steel was in the expected range and this supports the general idea. The influence of a relative thick skin depth should be investigated more precisely in subsequent investigations. Nevertheless, the method appears to be suitable for all metal types.

The parabolic behavior of piezo output voltage as a function of capacitor charge voltage (Fig. 6) is in good accordance with Eqs. (7) and (8) and other considerations in the introduction: ultrasound power ~ (electromagnetic power)^2. An effect of second order or quadratic behavior is present. Higher power levels should further increase efficiency and then resulting in much higher ultrasound intensities. Even just an increased *power density* can considerably improve the efficiency of the ultrasound transmission. It appears from Fig. 7 that for small lift off gaps (g: <10 mm) the smaller spiral coil with 1 cm^{2}is clearly superior for high intensity ultrasound. A small and quite powerful NDT transducer might be attractive for certain applications.

The internal inductances of the pulse generator must be held smaller than the inductive transducer for good power utilization (non-linear effect of electro-acoustic conversion!). The realization of higher frequencies towards 10 MHz probably requires smaller capacitors with smallest parasitic inductance. High energy content then must be achieved with even higher charge voltages.

For higher distances (cm) the geometries of the coil must be larger for better coupling. The achievable intensities however still strongly decline over distance.

The presented technique in the current state preferably transmits longitudinal waves and additionally, it cannot readily detect ultrasound. Ultrasound detection, therefore, must be realized by a separate and additional device and this could be a conventional EMAT or an optical receiver. Alternatively, the presented "coil only" transmitter can be equipped with an additional permanent magnet (only resulting in fields of typically < 1 T, Ref. [5]) and it then operates as a conventional EMAT for detection purposes. Another alternative would be the superimposition of a strong and short RF burst at relatively high frequencies (say 10 MHz) to the coil, at a time where the ultrasound echo is expected from target. An RF generator at quite different frequency (10 MHz) can be well decoupled from the 1 MHz pulsed power. The echo (2 MHz) mixes with the 10 MHz excitation (eddy currents at 10 MHz) and results into a 8 and 12 MHz electrical signal in the coil. The 12 MHz echo then could be separated and analyzed. As with a static field of conventional EMAT, the 10 MHz excitation should be strong for a better 12 MHz echo signal. But in contrast to a conventional EMAT, the 10 MHz field would be – again – concentrated by the skin effect and additionally it spatially well coincidences with the eddy current in the metal. And additionally, it preferably would detect longitudinal waves.

As another unsolved technical problem here, the small coil only transducer is not supplied over a flexible cable but directly connected to the relatively large pulse generator. The transmission of short pulses with many kA through a compact and flexible cable towards a small transmitter is certainly much more attractive for practical applications. This appears quite challenging and should be investigated in future work.

ACKNOWLEDGEMENT

This work is initiated and supported by the internal research founding of the University of Applied Sciences Ruhr-West.

REFERENCES

1. H.M. Frost, Electromagnetic-ultrasonic transducers: principles, practice and applications, in: W.P. Mason, E.N. Thurston (Eds.), Physical Acoustics, vol. XIV, Academic Press, New York, USA, 1979, p. 179.

2. R.B. Thompson, Physical principles of measurements with emat transducers, in: W.P. Mason, R.N. Thurston (Eds.), Physical Acoustics, vol. XIX, Academic Press, New York, 1990, pp. 157–200.

3. B.W. Maxfield, C.M. Fortunko, The design and use of electromagnetic acoustic wave transducers (emats), Mater. Eval. 41 (1983) 1399–1408.

4. W.P. Mason, R.N. Thurston, Physical Acoustics, Electromagnetic—Ultrasound Transducers: Principles, Practice and Applications, vol. XIV, Academic Press, New York, NY, 1970.

5. X. Jian, S. Dixon, R.S. Edwards, Modelling ultrasonic generation for Lorentz force EMATs, Insight (BINDT) 46 (11) (2004) 671–673.

6. X. Jian, S. Dixon, R.S. Edwards, J. Morrison, Coupling mechanism of an EMAT, Ultrasonics 44 (2006) e653–e656.

7. Josef Krautkrämer, Herbert Krautkrämer, Ultrasonic testing of materials, 4th fully rev. ed. Springer-Verlag, Berlin, New York, 1990. ISBN 3-540-51231-4.

8. G. Harvey, D.E. Brower, Metal Forming Device and Method, U.S. Patent Documents 2,976,907, General Dynamics Corporation, 28.08.1958.

9. B.P. Leftheris, Method of Welding Metals Using Stress Waves, U.S. Patent Documents 3,961,739, Grumman Aerospace Corporation, 07.05.1974.

10. R. Weadock, Magnetic Forming Apparatus, U.S. Patent Documents 3,348,397, General Motors Corporation, 22.10.1964.

11. Journal of Materials Processing Technology, vol. 211, Issue 5, Special Issue: Impulse Forming, in: A. Erman Tekkaya, Julian M. Allwood (Eds.) Electromagnetic Forming—A review, 1 May 2011, pp. 787–829.

12. See e.g. Wikipedia.org ‘‘skin effect’’ or standard text book electrodynamics.

13. See e.g. Wikipedia.org ‘‘magnetic pressure’’ or standard text book electrodynamics.

14. E.R. Dobbs, Electromagnetic generation of ultrasonic waves, in: W.P. Mason, R.N. Thurston (Eds.), Physical Acoustics, vol. X, Academic Press, New York, 1973, pp. 127–189.

15. E.R. Dobbs, J.D. Llewellyn, Non-destruct, Test 4 (1971) 49–56.

16. D. Basting, G. Marowsky, Excimer Laser Technology, Springer, 05.12.2005, ISBN: 3-540-20056-8.

17. E.E. Kunhardt, The formation of pulsed discharges, Proc. Fourth IEEE Pulsed Power Conf. (1983) 206–211.

18. C.B. Scruby, Some applications of laser ultrasound, Ultrasonics 27 (1989) 195– 209.

19. C.B. Scruby, R.J. Dewhurst, D.A. Hutchins, S.B. Palmer, Research Techniques in Non-destructive Testing, vol. 5, in: R.S. Sharpe (Ed.), Academic Press, London, UK, 1982, p. 281.

20. R.D. Huber, R.E. Green Jr., Non-contact acousto ultrasonics using laser generation and laser interferometric detection, Mater. Eval. 49 (1991) 613–618.

21. J.R. Hutchinson, C.M. Percival, Higher modes of longitudinal wave propagation in thin rods, J. Acoust. Soc. Amer. 44 (5) (1968) 1204–1210.

CHAPTER 3

On the scattering of elastic waves from a non-axisymmetric defect in a coated pipe

Wenbo Duan, Ray Kirby, Peter Mudge

HIGHLIGHTS

- A numerical method is proposed to study wave scattering from defects in coated pipes.
- Predictions are generated in the time and frequency domain.
- The longitudinal family of modes is likely to propagate longer than torsional modes.
- The coating does not induce any additional mode conversion.
- The effect of coating on modal dispersion and scattering can be neglected.

ABSTRACT

Viscoelastic coatings are often used to protect pipelines in the oil and gas industry. However, over time defects and areas of corrosion often form in these pipelines and so it is desirable to monitor the structural integrity of these coated pipes using techniques similar to those used on uncoated pipelines. A common approach is to use ultrasonic guided waves that work on the pulse-echo principle; however, the energy in the guided waves can be heavily attenuated by the coating and so significantly reduce the effective range of these techniques. Accordingly, it is desirable to develop a better understanding of how these waves propagate in coated pipes with a view to optimising test methodologies, and so this article uses a hybrid SAFE-finite element approach to model scattering from non-axisymmetric defects in coated pipes. Predictions are generated in the time and frequency domain and it is shown that the longitudinal family of modes is likely to have a longer range in coated pipes when compared to torsional modes. Moreover, it is observed that the energy velocity of modes in a coated pipe is very similar to the group velocity of equivalent modes in uncoated pipes. It is also observed that the coating does not induce any additional mode conversion over and above that seen for an uncoated pipe when an incident wave is scattered by a defect. Accordingly, it is shown that when studying coated pipes one need account only for the attenuation imparted by the coating so that one may normally neglect the effect of coating on modal dispersion and scattering.

KEYWORDS

Guided wave; Non-axisymmetric defect; Hybrid finite element method; Dispersion and scattering analysis

1. INTRODUCTION

Viscoelastic materials are often used as coatings on the outer surface of pipelines in order to protect the pipe from external damage and corrosion. However, over time it is possible for these coatings to degrade and for regions of corrosion or other defects to form within the pipe substrate. Accordingly, it is desirable to monitor the integrity of the pipe and a fast and efficient way to do this is through the use of non-destructive testing (NDT) techniques such as long range ultrasonic testing (LRUT) [1] and [2]. The application of LRUT to coated pipes involves sending a guided wave along the pipe wall, but this technique is less successful for coated pipes because the viscoelastic coating attenuates the ultrasonic wave as it travels along the pipe wall. This has the effect of significantly reducing the range over which LRUT can be successfully used in the location of defects such as corrosion. This presents a significant problem because LRUT is an important tool for interrogating pipelines and the use of viscoelastic coatings is relatively widespread. It is desirable, therefore, to try and develop a better understanding of the way in which a coating attenuates an ultrasonic wave, as well as how it affects the scattering of waves from defects. One approach to achieving a better understanding is through the development of theoretical models, however there are very few articles in the literature that use theoretical models to analyse scattering from defects in a pipeline coated with a viscoelastic material. Accordingly, this article utilises a three dimensional model that is suitable for analysing scattering from defects of arbitrary shape in a pipe of arbitrary length coated with a viscoelastic material. In doing so this model moves away from relying on dispersion curves in order to analyse a scattering problem that is more representative of problems found in the field.

A typical pipeline consists of a long and uniform pipe in which the defect, or region of corrosion, forms only over a short section of the pipe. LRUT works by sending an incident pulse down the pipe and then recording the reflected pulse scattered by the defect. The energy contained within the incident and reflected pulse travels as a series of eigenmodes and so one must be careful to excite the appropriate mode, or modes, as well as to retain some understanding of the characteristics of each mode when interpreting the returning pulse, and here a knowledge of group velocity is important when distinguishing modal content. Therefore, understanding the properties of the pipe eigenmodes is very important in the practical application of LRUT and so a common starting point for a theoretical analysis of coated pipes is to find the pipe eigenmodes, or dispersion curves as they are known. This approach is illustrated for coated pipes by Barshinger and Rose [3], who applied a global transfer matrix method to compute the phase and attenuation of axisymmetric longitudinal modes. The global matrix method derives an analytic expression for the governing dispersion relation and so numerical routines are necessary to find the complex roots of this equation. Root finding in the complex plane is often difficult and time consuming, especially at higher frequencies [3]and so it is advantageous to use alternative methods. Accordingly, numerical techniques are becoming increasingly popular and one of the most reliable and efficient methods for obtaining the eigenmodes of a uniform structure is the semi analytic finite element (SAFE) method. This approach substitutes an analytic expression for the displacement in the axial direction into Navier's governing equation and then uses the finite element method to solve the resulting two dimensional eigenequation. Thus, one only needs to mesh the cross section of the structure, and the SAFE method may be applied to structures with an arbitrary cross-section provided they are uniform in the axial direction. A rigorous introduction to the SAFE method is provided by Bartoli et al. [4], who proceed to apply the method to a waveguide of arbitrary cross-section, as well as a viscoelastic plate; for other examples of the application of the SAFE method see [5], [6], [7],[8] and [9]. Mu and Rose [10] also applied the SAFE method to pipes with a viscoelastic coating, and by using an analytic expansion for the circumferential direction they were able to further reduce the problem to one dimension. Moreover, through the use of an orthogonality relation for the pipe eigenmodes, Mu and Rose were able to sort values for phase velocity and attenuation for a large number of propagating eigenmodes. Further applications of the SAFE method to problems involving energy dissipation include the work of Castaings and Lowe [11], who calculate the eigenmodes for a waveguide of arbitrary cross-section that is surrounded by an absorbing region, and Marzani et al. [12]who examined multi-layered structures and computed the energy velocity for the eigenmodes, which is generally more appropriate than group velocity for structures in which material damping is present [13]. Thus, the SAFE method has now been shown to deliver a reliable and efficient means for finding the eigenmodes in a waveguide containing material damping and so this method is well suited to studying coated pipes. Accordingly, this article will make use of the SAFE method to calculate eigenmodes for uniform regions of a coated pipe and in the section that follows the SAFE method is applied to a two dimensional problem.

The SAFE method is very useful for finding the eigenmodes in an infinitely long structure, however the eigenexpansion assumes that the structure is uniform. If one is also to model the scattering from a defect then this adds considerable additional complexity, especially if one also wishes to study a non-axisymmetric defect. Modelling difficulties are caused by the non-uniformities in the structure and whilst it is possible simply to numerically discretise an entire pipe this likely to require extremely high numbers of degrees of freedom even for modest pipe lengths. Moreover, discretising the entire pipe obviously cannot be achieved if the pipe is infinite, and so this approach normally requires some form of non-reflecting boundary in order to close the problem for an equivalent finite length of pipe. An example of the finite element method was presented by Hua and Rose [14], who studied a short length of coated pipe. Hua and Rose used commercial software and studied the attenuation of guided waves in a uniform pipe where there is no additional mode conversion at the end of the pipe, however this method will quickly generate excessive degrees of freedom if one moves to more representative geometries. Predoi et al. [15] used an absorbing boundary layer method to study scattering from a defect in a two dimensional viscoelastic plate and it is clear that extension to three dimensions is likely to become computationally expensive. It is also possible to reduce computational expenditure by introducing higher order finite elements and Żak [16] demonstrates the application of the spectral element method to wave propagation in a plate. The spectral finite element is now well developed for structural health monitoring and the increase in computational efficiency that this provides means that it is now capable of being applied to relatively large structures [17], however one must still mesh the entire structure and this is not always the most attractive option, especially for structures such as pipelines that are long and slender. Moreover, structures such as pipelines also have a relatively simple geometry and it is possible to take advantage of this when developing a numerical model. This is achieved by using alternative methods for modelling wave propagation in the long uniform sections found in pipe installations, or similar guided wave applications. For example, Galán and Abascal[18] used a hybrid boundary element-finite element approach to study scattering from a defect in a plate coated with a viscoelastic material; this approach then reduces the degrees of freedom required in the uniform section through the use of a boundary element discretisation. An alternative approach that is potentially even more efficient is to use a modal expansion for the uniform section of pipe and to couple this to a numerical discretisation that surrounds only the defect being studied. This will radically reduce the number of degrees of freedom required when compared to a full discretisation and in principle it can be used for any length of pipe without incurring additional computational costs. A relevant example of this approach for uncoated pipes is the method of Zhou et al.[19], who used the wave finite element method to solve the eigenequation in the uniform section, and then coupled this to a finite element discretisation surrounding the defect. Recently Benmeddour et al. [20] used a hybrid SAFE-finite element (FE) method to analyse elastic wave propagation in a solid cylinder, and Duan and Kirby [21] used a similar method to analyse elastic wave propagation in an uncoated pipe. The article by Duan and Kirby contains a more detailed discussion about these and other alternative numerical approaches and so these will not be discussed further here. However, it is noticeable that in the literature the only application of a hybrid SAFE-FE approach to coated pipes was that reported by Kirby et al. [22] and [23]. The models developed by Kirby et al. were used primarily to deduce the bulk shear and longitudinal properties of the viscoelastic coating, and to do this it was necessary only to study the axisymmetric problem. Therefore, Kirby et al. restricted their analysis to either torsional [22] or longitudinal [23] modes, and so these approaches are not suitable for studying the more general problem of scattering from non-axisymmetric defects.

The aim of this article is to analyse scattering from non-axisymmetric defects in coated pipes using the hybrid SAFE-FE method. Relevant examples of the application of this method to elastic wave propagation in circular geometries include the articles by Benmeddour et al. [20], and Duan and Kirby [21]. Note that Benmeddour et al. use a variational formulation, whilst Duan and Kirby use a weighted residual formulation to derive the final system of equations. Thus, in Section 2 a weighted residual formulation is adopted for the three dimensional problem. In Section 3 predictions generated using a three dimensional model are validated against two dimensional predictions and measurements. In Sections 4 and 5 predictions are generated that quantify scattering from a non-axisymmetric defect representative of corrosion in a coated pipe, and here predictions are presented in the frequency and time domain. Parametric studies are also undertaken and conclusions drawn regarding the influence of the coating and on the choice of excitation when undertaking LRUT in coated pipes.

2. THEORY

The theory reported in this section is based on the hybrid SAFE-FE method described by Duan and Kirby [21]. However, the analysis of a coated pipe requires the addition of an extra layer when compared to the analysis reported by Duan and Kirby [21]. Accordingly, when adding an additional layer it is convenient to use a normalisation procedure that is different to that used by Duan and Kirby, as this facilitates the writing of the final governing equations in a computationally efficient way. The governing equation for wave propagation in an elastic or viscoelastic medium is Navier's equation, which is written as

$$\left(\lambda_{\mathrm{p,c}} + \mu_{\mathrm{p,c}}\right)\nabla\left(\nabla \cdot \boldsymbol{u}'_{\mathrm{p,c}}\right) + \mu_{\mathrm{p,c}}\nabla^2\boldsymbol{u}'_{\mathrm{p,c}} = \rho_{\mathrm{p,c}}\frac{\partial^2 \boldsymbol{u}'_{\mathrm{p,c}}}{\partial t^2}, \tag{1}$$

where λ and μ are the Lamé constants, $\boldsymbol{u}'$ is the displacement vector, ρ is density and t is time. A time dependence of $e^{i\omega t}$ is assumed throughout this article, where ω is the radian frequency and $i = \sqrt{-1}$. The subscripts p and c denote pipe and coating respectively. Navier's equation is applied here to an infinite pipe coated with a viscoelastic material, see Fig. 1, with the coating applied to the outer surface. The pipe also contains a single defect that is arbitrary in shape and which penetrates both the coating and the pipe substrate. The boundary condition for this problem is assumed to be traction free on the outer surface of the coating and the inner surface of the pipe. On the interface between the pipe and the coating, the displacement and traction forces are equal. A hybrid SAFE-finite element approach is adopted so that a numerical discretisation is used for a region surrounding the defect (this is Ω_2 in Fig. 1), and for those uniform sections abutting this region a modal expansion is adopted (regions Ω_1 and Ω_3). Thus, for the uniform sections the SAFE method is applied to find the pipe eigenmodes and this is discussed in the following section; a full finite element discretisation of Ω_2 is then discussed in Section 2.2.

2.1. SAFE method for a coated pipe

The displacements u'_{1q} in region Ω_1 of the pipe are expanded over the pipe eigenmodes to give

$$u'_{1q}(x, y, z) = \sum_{n=0}^{\infty} u^n_{1q}(x, y)e^{-ik_\gamma^n z}, \tag{2}$$

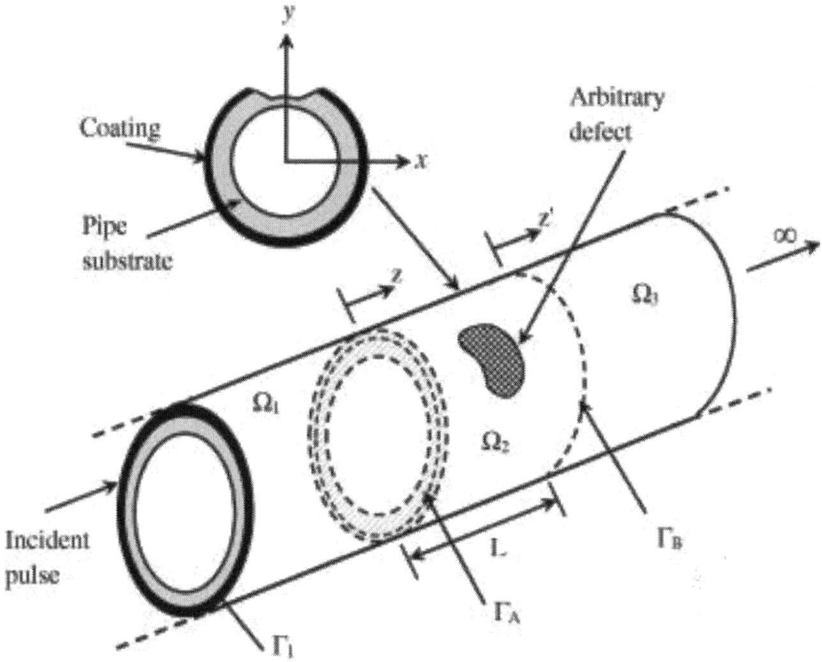

Figure. 1. Geometry of coated pipe containing arbitrary defect.

where the subscript $q = x, y$ or z, and $u_{1q}(x,y)$ are the eigenvectors in region Ω_1, where $k=\omega/c_{Tp}$ so that γ is a (coupled) dimensionless wavenumber. In addition,c_{Tp} and c_{Lp} are the shear (torsional) and compressional (longitudinal) bulk wave velocities in the pipe substrate, respectively. Further, c_{Tc} and c_{Lc} are the shear (torsional) and compressional (longitudinal) bulk wave velocities in the viscoelastic coating, respectively. The finite element analysis proceeds by discretising the displacements of any mode n over the pipe cross-section to give

$$u_{1q}(x,y) = \sum_{j=1}^{p_{1q}} N_{qj}(x,y) u_{1qj} = \mathbf{N}_q \mathbf{u}_{1q}, \tag{3}$$

where N_{qj} is a global trial (or shape) function, u_{1qj} is the value of u_{1q} at node j , andp_{1q} is the number of nodes (or degrees of freedom) for the displacements in direction q. In addition, $\mathbf{N}_q$ and $\mathbf{u}_{1q}$ are row and column vectors of length p_{1q}, respectively, and it is convenient to choose $\mathbf{N}_x=\mathbf{N}_y=\mathbf{N}_z$ so that one only needs to generate one finite element mesh. The displacement can be further divided as $\mathbf{u}_{1q}=[\mathbf{u}_{1qp}\mathbf{u}_{1qc}]^T$, where $\mathbf{u}_{1qp}$ and $\mathbf{u}_{1qc}$ denote nodal displacements in the pipe and in the coating, respectively. This also facilitates the application of continuity of displacement and traction forces over the interface between the pipe and the coating, and after applying these continuity conditions and enforcing zero traction on the internal surface of the pipe and the outer edge of the coating [21], the substitution of Eq. (3) into Eq. (1) yields the following eigenequation:

$$\mathbf{P}\mathbf{u}_1=\gamma\mathbf{S}\mathbf{u}_1, \tag{4}$$

where $\mathbf{u}_1 = [\mathbf{u}_{1x} \ \ \mathbf{u}_{1y} \ \ \mathbf{u}_{1z} \ \ \gamma\mathbf{u}_{1x} \ \ \gamma\mathbf{u}_{1y} \ \ \gamma\mathbf{u}_{1z}]^T$. It is convenient to split up matrices $\mathbf{P}$ and $\mathbf{S}$ so that

$$\mathbf{P}\mathbf{u}_1=\mathbf{P}_p\mathbf{u}_{1p}+\mathbf{P}_c\mathbf{u}_{1c} \tag{5}$$

$$\mathbf{S}\mathbf{u}_1=\mathbf{S}_p\mathbf{u}_{1p}+\mathbf{S}_c\mathbf{u}_{1c} \tag{6}$$

where $\mathbf{u}_{1p,c} = [\mathbf{u}_{1xp,c} \ \ \mathbf{u}_{1yp,c} \ \ \mathbf{u}_{1zp,c} \ \ \gamma\mathbf{u}_{1xp,c} \ \ \gamma\mathbf{u}_{1yp,c} \ \ \gamma\mathbf{u}_{1zp,c}]^T$, and the constituents of matrices $\mathbf{P}_{p,c}$ and $\mathbf{S}_{p,c}$ are given in Appendix A. Eq. (4) is a sparse symmetric eigenequation that is solved for the eigenmodes in the coated pipe. The method for solving this equation and for sorting the eigenmodes that are obtained is described by Duan and Kirby [21]. This solution delivers $p_1=p_3=p_{1x}+p_{1y}+p_{1z}$ eigenmodes.

Modal attenuation and energy velocity are two of the most important factors to be considered when examining wave propagation in a coated pipe. Attenuation quantifies the reduction in amplitude of the wave, whereas the energy velocity determines the velocity of the "centre of gravity" of a wave as it propagates along the pipe wall [24]. For attenuative waves in coated pipes, energy velocity more accurately represents the velocity of a wave package, although of course the energy velocity reduces to the group velocity for an uncoated pipe. The attenuation (Δ) is defined in the usual way:

Δ=-20I(k)×log$_{10}$(e). (7)

The energy velocity is defined as the ratio of the average power flow to the average stored energy per unit length of the waveguide [24] to give

$$V_e = \frac{2Re\left\{\sum_{\tau=p,c}\int_{\Gamma_{A\tau}}\left[-\sigma_{1zq-}\cdot u^*_{1q-}\right]d\Gamma_{A\tau}\right\}}{Re\left\{\sum_{\tau=p,c}\int_{\Gamma_{A\tau}}\left[\rho_\tau u_{1q-}\cdot u^*_{1q-}+\sigma_{1ql-}:\varepsilon^*_{1ql-}\right]d\Gamma_{A\tau}\right\}}. \quad (8)$$

Here, the superscript $*$ indicates the complex conjugate, and q and l take on values of x,y and z, with the summation convention applying to repeated indices of q and l . In addition, u_{1q-},σ_{1ql-} and ε_{1ql-} are the incident displacements, stress and strain tensors in region Ω_1. The first term in the denominator represents the kinetic energy density and the second term the strain energy density.

2.2. A hybrid SAFE-FE approach for a coated pipe

A three dimensional finite element discretisation is used for the non-uniform section of the pipe, Ω_2, which is assumed to contain a defect of arbitrary shape, see Fig. 1. The displacements $u'_{2q}(x,y,z)$ in Ω_2 are discretised to give

$$u'_{2q}(x,y,z)=\sum_{j=1}^{p_{2q}}W_{qj}(x,y,z)u_{2qj}=\mathbf{W}_q\mathbf{u}_{2q}, \quad (9)$$

where W_{qj} is a global shape function, and u_{2qj} is the value of u'_{2q} at node j , and $p_{2q}$is the number of nodes in the q direction. For convenience when matching with the inlet and outlet regions, we choose $\mathbf{W}_x=\mathbf{W}_y=\mathbf{W}_z=\mathbf{W}$ so that the weak forms of Eq.(1) yields

$$\int_{\Omega_2}\left\{\left[\left(\lambda_{p,c}+\mu_{p,c}\right)\frac{\partial\mathbf{W}^T}{\partial x}\frac{\partial\mathbf{W}}{\partial x}+\mu_{p,c}\nabla\mathbf{W}^T\nabla\mathbf{W}-\rho\omega^2\mathbf{W}^T\mathbf{W}\right]\mathbf{u}_{2xp,c}\right.$$
$$+\left[\lambda_{p,c}\frac{\partial\mathbf{W}^T}{\partial x}\frac{\partial\mathbf{W}}{\partial y}+\mu_{p,c}\frac{\partial\mathbf{W}^T}{\partial y}\frac{\partial\mathbf{W}}{\partial x}\right]\mathbf{u}_{2yp,c}$$
$$\left.+\left[\lambda_{p,c}\frac{\partial\mathbf{W}^T}{\partial x}\frac{\partial\mathbf{W}}{\partial z}+\mu_{p,c}\frac{\partial\mathbf{W}^T}{\partial z}\frac{\partial\mathbf{W}}{\partial x}\right]\mathbf{u}_{2zp,c}\right\}d\Omega_2=\int_{\Gamma_2}\mathbf{W}^T h_{2xp,c}d\Gamma_2,$$

(10a)

$$\int_{\Omega_2}\left\{\left[\lambda_{p,c}\frac{\partial\mathbf{W}^T}{\partial y}\frac{\partial\mathbf{W}}{\partial x}+\mu_{p,c}\frac{\partial\mathbf{W}^T}{\partial x}\frac{\partial\mathbf{W}}{\partial y}\right]\mathbf{u}_{2xp,c}+\left[\left(\lambda_{p,c}+\mu_{p,c}\right)\frac{\partial\mathbf{W}^T}{\partial y}\frac{\partial\mathbf{W}}{\partial y}\right.\right.$$
$$\left.+\mu_{p,c}\nabla\mathbf{W}^T\nabla\mathbf{W}-\rho\omega^2\mathbf{W}^T\mathbf{W}\right]\mathbf{u}_{2yp,c}$$
$$\left.+\left[\lambda_{p,c}\frac{\partial\mathbf{W}^T}{\partial y}\frac{\partial\mathbf{W}}{\partial z}+\mu_{p,c}\frac{\partial\mathbf{W}^T}{\partial z}\frac{\partial\mathbf{W}}{\partial y}\right]\mathbf{u}_{2zp,c}\right\}d\Omega_2=\int_{\Gamma_2}\mathbf{W}^T h_{2yp,c}d\Gamma_2,$$

(10b)

$$\int_{\Omega_2}\left\{\left[\lambda_{p,c}\frac{\partial\mathbf{W}^T}{\partial z}\frac{\partial\mathbf{W}}{\partial x}+\mu_{p,c}\frac{\partial\mathbf{W}^T}{\partial x}\frac{\partial\mathbf{W}}{\partial z}\right]\mathbf{u}_{2xp,c}+\left[\lambda_{p,c}\frac{\partial\mathbf{W}^T}{\partial z}\frac{\partial\mathbf{W}}{\partial y}\right.\right.$$
$$\left.+\mu_{p,c}\frac{\partial\mathbf{W}^T}{\partial y}\frac{\partial\mathbf{W}}{\partial z}\right]\mathbf{u}_{2yp,c}+\left[\left(\lambda_{p,c}+\mu_{p,c}\right)\frac{\partial\mathbf{W}^T}{\partial z}\frac{\partial\mathbf{W}}{\partial z}+\mu_{p,c}\nabla\mathbf{W}^T\nabla\mathbf{W}\right.$$
$$\left.\left.-\rho\omega^2\mathbf{W}^T\mathbf{W}\right]\mathbf{u}_{2zp,c}\right\}d\Omega_2=\int_{\Gamma_2}\mathbf{W}^T h_{2zp,c}d\Gamma_2.$$

(10c)

Here, the notation for the pipe substrate and coating regions follows the conventions used in the previous section. Similarly, it is necessary also to enforce continuity of displacement and traction force between the coating and the pipe, as well as enforce zero traction force over the outer surface of Ω_2, apart from the surfaces Γ_A and Γ_B, so that

$$h_{2qp,c}=\sigma_{2ql_{p,c}}n_l=0, \tag{11}$$

where the indices q and l take on values of x,y and z, and the summation convention applies to repeated indices of l only. In Eq. (11), σql denotes the Cauchy stress tensor and n_l is the unit outward normal vector to the surface of the pipe so that

$$h_{xp,c} = \lambda_{p,c}\left(\frac{\partial u'_{2xp,c}}{\partial x}+\frac{\partial u'_{2yp,c}}{\partial y}+\frac{\partial u'_{2zp,c}}{\partial z}\right)n_x + 2\mu_{p,c}\frac{\partial u'_{2xp,c}}{\partial x}n_x + \mu_{p,c}\left(\frac{\partial u'_{2xp,c}}{\partial y}+\frac{\partial u'_{2yp,c}}{\partial x}\right)n_y + \mu_{p,c}\left(\frac{\partial u'_{2xp,c}}{\partial z}+\frac{\partial u'_{2zp,c}}{\partial x}\right)n_z, \tag{12a}$$

$$h_{yp,c} = \mu_{p,c}\left(\frac{\partial u'_{2yp,c}}{\partial x}+\frac{\partial u'_{2xp,c}}{\partial y}\right)n_x + \lambda_{p,c}\left(\frac{\partial u'_{2xp,c}}{\partial x}+\frac{\partial u'_{2yp,c}}{\partial y}+\frac{\partial u'_{2zp,c}}{\partial z}\right)n_y + 2\mu_{p,c}\frac{\partial u'_{2yp,c}}{\partial y}n_y + \mu_{p,c}\left(\frac{\partial u'_{2yp,c}}{\partial z}+\frac{\partial u'_{2zp,c}}{\partial y}\right)n_z, \tag{12b}$$

$$h_{zp,c} = \mu_{p,c}\left(\frac{\partial u'_{2zp,c}}{\partial x}+\frac{\partial u'_{2xp,c}}{\partial z}\right)n_x + \mu_{p,c}\left(\frac{\partial u'_{2zp,c}}{\partial y}+\frac{\partial u'_{2yp,c}}{\partial z}\right)n_y + \lambda_{p,c}\left(\frac{\partial u'_{2xp,c}}{\partial x}+\frac{\partial u'_{2yp,c}}{\partial y}+\frac{\partial u'_{2zp,c}}{\partial z}\right)n_z + 2\mu_{p,c}\frac{\partial u'_{2zp,c}}{\partial z}n_z. \tag{12c}$$

Continuity of displacement and traction force between the coating and the pipe is implemented naturally in region 2 once the region is discretised appropriately.

The displacements in regions Ω_1 and Ω_3 are written as modal expansions using solutions from the previous section, so that

$$u'_{1q}(x,y,z) = \sum_{n=0}^{m_1} A^n u^n_{1q+}(x,y)e^{-ik_\gamma^n z} + \sum_{n=0}^{m_1} B^n u^n_{1q-}(x,y)e^{ik_\gamma^n z}, \tag{13}$$

and

$$u'_{3q}(x,y,z') = \sum_{n=0}^{m_1} C^n u^n_{1q+}(x,y)e^{-ik_\gamma^n z'}. \tag{14}$$

Here, A^n,B^n and C^n are modal amplitudes, and u^n_{1q+} and u^n_{1q-} are eigenvectors for the incident and reflected waves, respectively. The number of modes used in the analysis for regions Ω_1 and Ω_3 is m_1, where $m_1 \leqslant p_1$. It is assumed that the pipe extends to infinity in region Ω_3 so that no reflected waves are present in this region. In addition, Eq. (13) allows for a general incident sound field, although in the analysis that follows this will be restricted either to torsional T(0,1) or longitudinal L(0,2)excitation as this best reflects experimental practice.

The problem is solved by enforcing continuity of displacement and traction forces in the axial direction over the interface between the uniform pipe sections and the central finite element based solution. Each continuity condition is weighted using the method described by Duan and Kirby in [21] and this delivers a final system equation of the form:

$$\begin{bmatrix} -\mathbf{G}_{11-} & \mathbf{G}_{21}^{T} & \mathbf{G}_{31}^{T} & \mathbf{G}_{41-}^{T} & \mathbf{0} \\ \mathbf{G}_{21} & \mathbf{G}_{22} & \mathbf{G}_{32}^{T} & \mathbf{G}_{42}^{T} & \mathbf{G}_{25} \\ \mathbf{G}_{31} & \mathbf{G}_{32} & \mathbf{G}_{33} & \mathbf{G}_{43}^{T} & \mathbf{G}_{35} \\ \mathbf{G}_{41-} & \mathbf{G}_{42} & \mathbf{G}_{43} & \mathbf{G}_{44} & \mathbf{G}_{45} \\ \mathbf{0} & \mathbf{G}_{25}^{T} & \mathbf{G}_{35}^{T} & \mathbf{G}_{45}^{T} & \mathbf{G}_{55} \end{bmatrix} \begin{Bmatrix} \mathbf{B} \\ \mathbf{u}_{2x} \\ \mathbf{u}_{2y} \\ \mathbf{u}_{2z} \\ \mathbf{C} \end{Bmatrix} = \begin{Bmatrix} \mathbf{G}_{11+}\mathbf{A} \\ \rho c_s^2 \left[\mathbf{Q}_{1x+} - \mathbf{Q}_{1zx+}\right]\mathbf{A} \\ \rho c_s^2 \left[\mathbf{Q}_{1y+} - \mathbf{Q}_{1zy+}\right]\mathbf{A} \\ -\mathbf{G}_{41+}\mathbf{A} \\ \mathbf{0} \end{Bmatrix}. \tag{15}$$

Eq. (15) is a set of $n_t\ (= 2m_1 + p_2)$ linear equations, where p_2 is the number of degrees of freedom in region Ω_2 and m_1 is the number of modes in regions Ω_1 andΩ_3 respectively. The vectors **A**, **B** and **C** hold the modal amplitudes A^n,B^n and C^nrespectively, and the other matrices are given in Appendix B. The modal amplitudes inΩ_1 and Ω_3, and the displacements in region Ω_2 are therefore found on the solution of Eq. (15). The displacements in Ω_1 and Ω_3 are then obtained from Eqs. (13) and (14).

The modal amplitudes defined in Eqs. (13) and (14) are independent of location in regionΩ_1. However, the displacement amplitude is dependent on location because the coating attenuates energy. Therefore, in order to study wave scattering from a defect without the influence of location, the reflection coefficient is defined here using a plane that is immediately adjacent to the left hand side of the defect. Thus, if we denote the distance between plane Γ_A and the left side of the defect as z_l then the reflection coefficient Λ for mode (m,n) [in the frequency domain only] is defined as

$$\Lambda^{(m,n)} = \frac{\left| B^{(m,n)} u_{1q-}^{(m,n)} e^{ik_\gamma^{(m,n)} z_l} \right|}{\left| A^{(m_I,n_I)} u_{1q+}^{(m_I,n_I)} e^{-ik_\gamma^{(m,n)} z_l} \right|}, \quad q = \theta \text{ or } z. \tag{16}$$

The reflection coefficient Λ is thus defined by normalising a reflected mode (m,n) by an incident mode (m_I,n_I), where the incident mode is determined by the choice of excitation, which will be either torsional (θ) or longitudinal (z). That is, if the incident mode is T(0,1) then $q=\theta$, and if it is L(0,2) then q=z. Building in the appropriate torsional or longitudinal displacements into Eq. (16) thus permits the comparison with measurements taken with T(0,1) and L(0,2) incident modes.

3. MODEL VALIDATION

The three dimensional model is validated here by comparing predictions against those available for two dimensional problems. Accordingly, the SAFE method is validated first, and then predictions obtained using the hybrid method for a coated pipe are compared against those in the literature for an axisymmetric defect. In the validation that follows, as well as the results presented in following sections, six noded triangular isoparametric elements are used for planes Γ_A and Γ_B, and ten noded tetrahedral isoparametric elements are used for Ω_2. The numerical model is programmed in MATLAB and executed on a computer with 12 CPU cores and a total accessible RAM of 128 GB. Parallelisation techniques are used so that the problem is solved in parallel for 12 frequencies at a time. Further, the properties of the coating follow the definition of Barshinger and Rose [3], so that the shear torsional and longitudinal bulk wave velocities are given as $c_{T_c} = 1/[1/\tilde{c}_T - i\tilde{\alpha}_T]$ and $c_{L_c} = 1/[1/\tilde{c}_L - i\tilde{\alpha}_L]$, respectively.

Here, $\tilde{c}_T$ and $\tilde{c}_L$denote the shear and longitudinal phase velocities, and $\tilde{\alpha}_T$ and $\tilde{\alpha}_L$ represent attenuation in the coating.

3.1. SAFE method for a coated pipe

The two dimensional SAFE method presented here is valid for a waveguide of any cross-sectional geometry, although in this article we restrict application to an axisymmetric pipe. A two dimensional solution is used as this makes the application of the hybrid method that follows more straightforward, especially when using flexural modes to enforce the matching conditions over Γ_A and Γ_B. Accordingly, the two dimensional SAFE solution is validated first by comparison against the independent one dimensional predictions for an axisymmetric pipe reported by Marzani et al. [12]. The pipe studied by Marzani et al. was filled with a viscoelastic material, rather than coated on the outside, and the pipe had an inner radius of 6.8 mm and an outer radius of 7.5 mm. The properties of the copper substrate are $c_{T_p} = 2240$ m/s, with a density $\rho_P = 8900$ kg/m^3. For the viscoelastic filling, $\tilde{c}_T = 430$ m/s,$\tilde{\alpha}_T = 0.5 \times 10^{-3}$ s/m and $\rho_C = 970$ kg/m^3. For the purposes of comparison, the two dimensional SAFE predictions are generated using 1288 nodes in the copper substrate and 3261 nodes in the viscoelastic filling, so that$p_{1x}=p_{1y}=p_{1z}=4357$ and the total degrees of freedom $p_1 = 13{,}071$. In Fig. 2predictions are compared against those found by Marzani et al. [12] for the *T*(0, 1) mode only, and it is seen that the current method delivers predictions that overlay those of Marzani et al. This provides evidence that the two dimensional SAFE method is working correctly, at least for this problem. Further validation was obtained by comparing predictions against the one dimensional model of Kirby et al. [22] and [23] and similar levels of agreement for longitudinal modes were also observed (not shown here), provided of course that sufficient nodes were included in the two dimensional model.

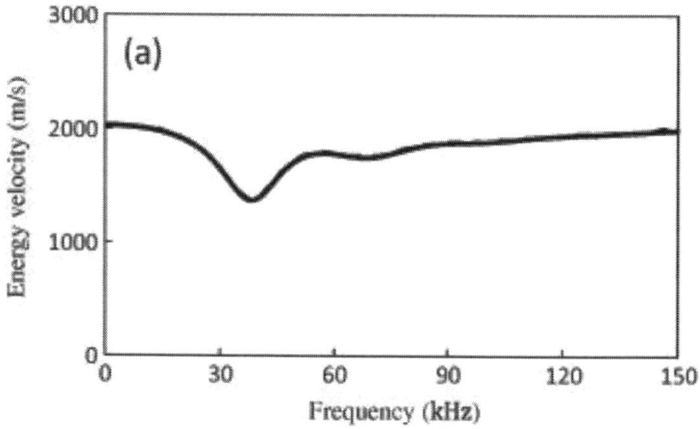

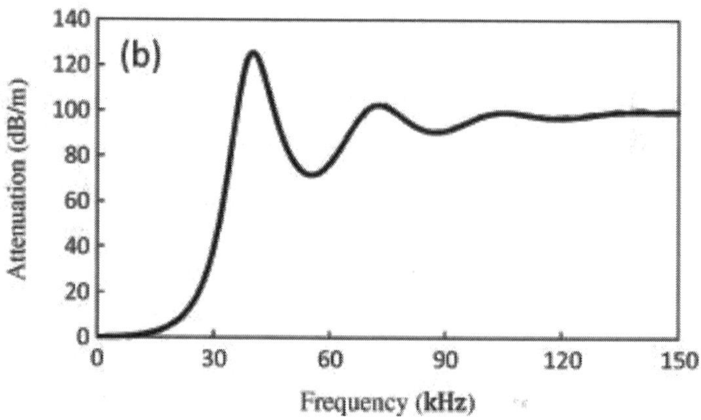

Figure. 2. Comparison between two dimensional SAFE predictions and those of Marzani et al. [12] for *T*(0, 1): —, current SAFE model, – – –, one dimensional SAFE model of Marzani et al.; (a) energy velocity, (b) attenuation; (solutions overlay one another).

3.2. Hybrid method for a coated pipe

There are far fewer results in the literature for pipes containing discontinuities and so validation against existing predictions is much more difficult for the hybrid method. The only suitable studies for coated pipes are the axisymmetric investigations of Kirby et al.[22] and [23], and so the three dimensional hybrid model is validated here by comparing predictions against a two dimensional model for an axisymmetric defect. Accordingly, the square defect used by Kirby et al. [22] and [23] is chosen, which is axisymmetric and uniform in the axial direction with a length of 15 mm, so that it cuts through the coating and penetrates 2.8 mm into the steel pipe substrate. The inner radius of the steel pipe is 39 mm and the outer radius of pipe substrate is 44.65 mm, with a coating thickness of 1.5 mm. The properties of the steel substrate are $c_{T_p} = 3260$ m/s, $c_{L_p} = 5960$ m/s, with a density $\rho_p = 8030$ kg/m^3. For the coating, $\tilde{c}_T = 750$ m/s, $\tilde{c}_L = 1860$ m/s, $\tilde{\alpha}_T = 3.9 \times 10^{-3}$ s/m and $\tilde{\alpha}_L = 0.023 \times 10^{-3}$ s/m, and $\rho_C = 1200$ kg/m^3. At each frequency, the two dimensional SAFE predictions for this problem are generated using 6680 nodes in the steel substrate and 3892 nodes in the viscoelastic coating, so that $p_{1x}=p_{1y}=p_{1z}=9820$ and $p_1 = 29{,}460$. The hybrid model then uses 200 of those modes found in the SAFE solution, so that $m_1=200$ and for the finite element mesh in region Ω_2,$p_2 = 489{,}228$.

In Fig. 3(a) and (b) the reflection coefficients for an axisymmetric defect are compared for excitation by T(0,1) and L(0,2) incident modes, respectively. The reflection coefficient used in these figures is the one defined in the article by Duan and Kirby [21]. It is seen in Fig. 3(a) and (b) that the agreement between the two and three dimensional approach is very good, and agreement was achieved to an accuracy of at least two decimal places over the frequency range shown. Of course, these two models should agree very well as they are solving the same problem, however the results in Fig. 3illustrate that it is possible to implement a three dimensional model and deliver accurate predictions over a wide frequency range in a reasonable computational time.

4. FREQUENCY DOMAIN: RESULTS AND DISCUSSION

The hybrid method reported in the previous sections is designed to allow a more detailed investigation into wave propagation in a coated pipe with a view to optimising the performance of commercial LRUT devices. The most significant problem encountered in experimental testing is that the coating attenuates the incident and reflected signal and so one finds it difficult to detect defects that are a relatively long distance from the source of excitation. However, it is not known if the coating significantly affects the way in which the waves are scattered by a defect, or how the coating affects the interpretation of the returning pulse in the time domain. Accordingly, the following sections address these questions and attempt to guide LRUT in coated pipes towards a more optimal approach. This is best achieved by first analysing the influence of the coating in the frequency domain, which is then followed in the next section by an investigation in the time domain. The frequency domain analysis that follows splits into two parts, the first examines the behaviour of the coated pipe eigenmodes, whereas the second part examines scattering from a defect. In the example that follows the coated section of the pipe has the following properties (unless otherwise specified): the pipe substrate is made from steel and the coating is the same viscoelastic material as that described in Section 3.2. The inner radius of the pipe is 39 mm, the outer radius of the steel substrate is 44.65 mm, and a reference coating with a thickness of 1.5 mm is chosen.

4.1. Mode shapes for a coated pipe

To gain an insight into the way in which each mode propagates in a coated pipe it is helpful to start by examining the shape of each eigenmode, and in particular the effect the coating has on each mode shape. Accordingly, in Fig. 4(a)–(c) the mode shape for a number of different modes is presented. Note that only a finite number of least attenuated modes are analysed here so that those modes with high levels of attenuation over the frequency range of interest in LRUT (roughly 20–100 kHz) are neglected. Moreover, inFig. 4 these modes are grouped together according to their mode shape so that three different families of modes are presented. Thus, in Fig. 4(a) the family containing $T(0, 1)$,$F(1, 2)$, $F(2, 2)$ and $F(3, 2)$ is presented [hereafter referred to as the $T(0, 1)$ family]; Fig. 4(b) contains $L(0, 2)$, $F(1, 3)$, $F(2, 3)$ and $F(3, 3)$ [the $L(0, 2)$ family]; and Fig. 4(c) containsL(0, 1), *F*(1, 1), *F*(2, 1) and *F*(3, 1) [the *L*(0, 1) family]. Note that at 70 kHz, *T*(0, 1) overlaysF(1, 2) and *L*(0, 1) overlays *F*(1, 1) in Fig. 4. The mode shapes illustrated in Fig. 4 are important in understanding how each mode behaves when a coating is added. This is most apparent with the large displacement drop seen within the coating for the circumferential eigenvector in Fig. 4(a). This has important ramifications for the use of torsional modes in LRUT because these modes require excitation in the circumferential direction. Furthermore, if one attaches a commercial LRUT device directly onto the outside of the coating and attempts to drive the *T*(0, 1) family, then the device is likely to find it difficult to transfer significant levels of energy into this family of modes and so there will be only a limited range over which the technique will work. Thus, if one insists on using the *T*(0, 1) family to find defects in coated pipes one must at least remove the coating before attaching the test equipment.

In Fig. 4(b) it is seen that for the *L*(0, 2) family the displacement drop in the coating region is smaller than that seen for the *T*(0, 1) family. It is significant that this is observed also forL(0, 2), which is normally the excitation mode of choice if one is using longitudinal modes, and this drop is not as severe as that seen for *T*(0, 1). Therefore, these figures illustrate that *L*(0, 2) is likely to be a better choice for LRUT when compared to *T*(0, 1), although the displacement drop for the *L*(0, 2) family is still significant and it appears prudent also to remove the coating when attempting to test sections of coated pipe. It is interesting to note that in Fig. 4(c) the *L*(0, 1) family is actually the best choice for the study of coated pipes. However, there are known to be practical difficulties when attempting to exciteL(0, 1), as it is harder to focus the energy into this mode when compared to *L*(0, 2). Moreover, in the frequency range of interest *L*(0, 1) is also more dispersive than *L*(0, 2) (see also Fig. 6(a) in Section 4.3) and so it is not normally favoured. Nevertheless, the results presented in Fig. 4 indicate that if these practical problems could be overcome, the *L*(0, 1) family is the most attractive alternative for studying scattering from defects in coated pipes.

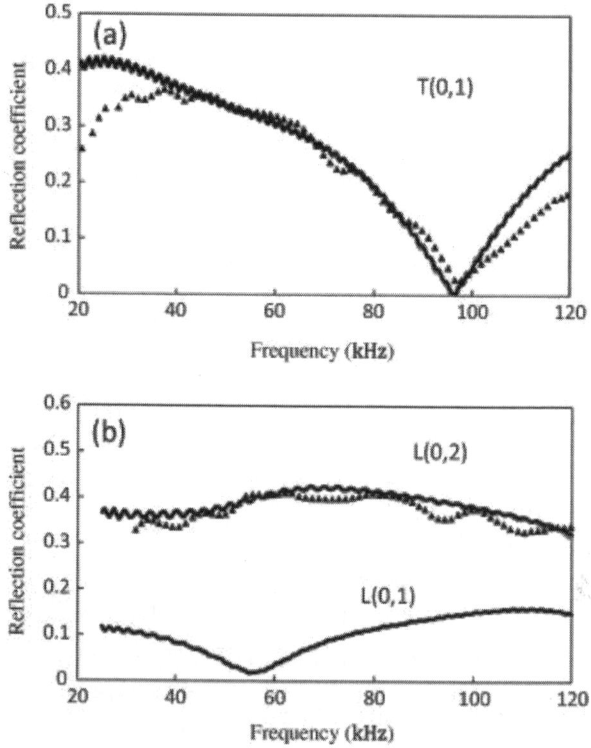

Figure. 3. Reflection coefficients for (a) the T(0,1) mode and (b) the L(0,2) mode incident upon an axisymmetric defect: —, current three dimensional model; – – –, two dimensional axisymmetric models of Kirby et al.[22] and [23]; ▲, Experiment [22] and [23].

4.2. Attenuation in a coated pipe

The influence of the coating on mode shape seen in Fig. 4 may also be observed in the relative attenuation of each mode, and this is shown in Fig. 5(a)–(d) for coating thicknesses of 1.5 mm, 3 mm and 5 mm. It can be seen in Fig. 5 that the attenuation of modes that sit within each family is similar to one another, especially as one moves away from the equivalent modal cut-on frequency for an uncoated pipe. This similarity in attenuation is caused by the similarity in mode shape seen in Fig. 4. It is interesting also to observe the effect of changing the coating thickness and here the *T*(0, 1) family is seen to be especially sensitive to coating thickness at lower frequencies. This behaviour indicates that the *T*(0, 1) family is potentially more sensitive to changes in material properties in the low frequency region, and this coincides with the frequency range where one may wish to attempt measurements on coated pipes (on the basis that one would expect lower levels of attenuation at lower frequencies). Moreover, the sensitivity of*T*(0, 1) is seen to be greater than that for *L*(0, 2), which exhibits consistently lower levels of attenuation, especially at low frequencies. This provides further evidence to support the use of longitudinal modes in LRUT on coated pipes.

4.3. Energy velocity in coated pipes

The relative attenuation of each mode is clearly very important when choosing the optimum strategy for LRUT in coated pipes. Another important quantity is the energy velocity of the wave as it travels along the pipe. The energy velocity is used extensively in uncoated pipes, where it is known as the group velocity. Energy velocity is used in the time domain to aid in separating different modes, as it enables one to calculate the time of flight of the propagating mode of interest. The energy velocity is defined in Eq. (8), and inFig. 6 the energy velocity with and without a coating is compared for the three modal families. To further investigate the effects of a coating, the energy velocity is also plotted for three different coating thicknesses, and here it is evident that the energy velocity of each mode remains largely unaffected by the presence of the coating, at least over the frequency range covered here. This demonstrates that the action of the coating is largely restricted to damping the energy carried by the wave and that it does not add significant levels of mass or stiffness to the system. Furthermore, one may see that *T*(0, 1) and*L*(0, 2) retain their non-dispersive characteristics over the frequency range of interest here, and so one may conclude that for LRUT the time of flight techniques used to separate modes in uncoated pipes are applicable also to coated pipes. Of course, the time domain response seen for a coated pipe will be different because of modal attenuation, and this is investigated in the next section.

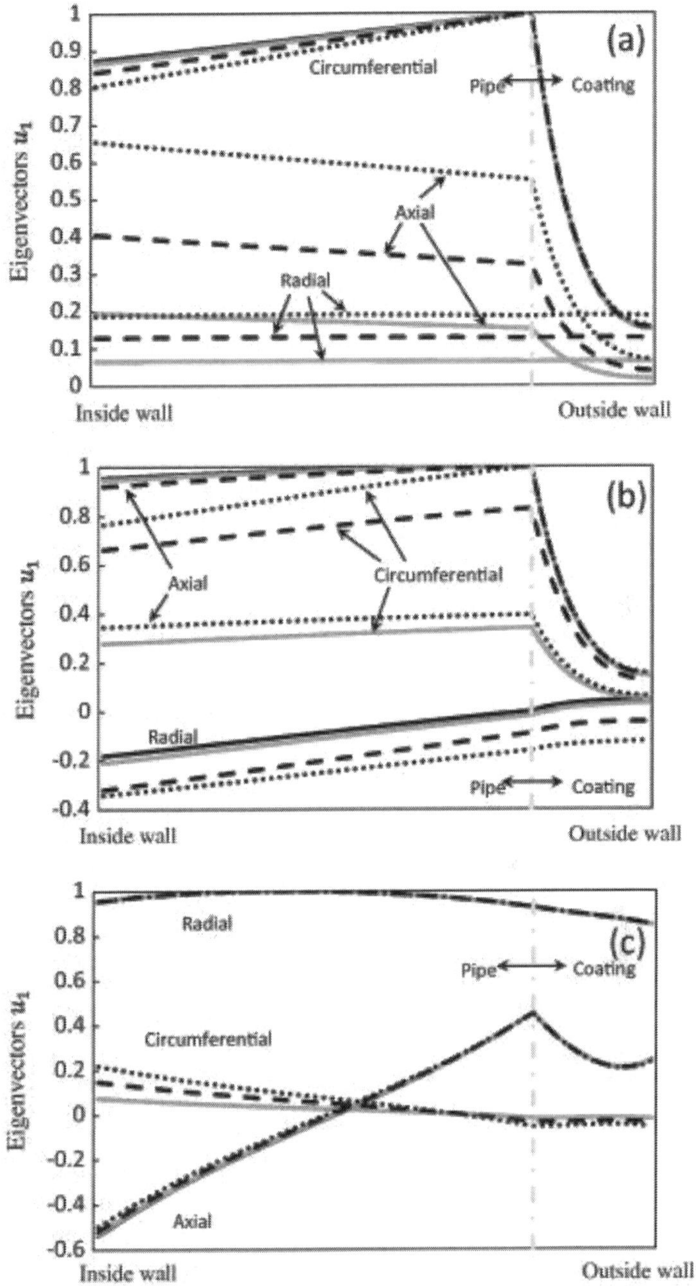

Figure. 4. Mode shapes for 3 in. schedule 40 coated pipe at 70 kHz. (a) —, $T(0, 1)$; —, $F(1, 2)$; – – –, $F(2, 2)$; ···,$F(3, 2)$. (b) —, $L(0, 2)$; —, $F(1, 3)$; – – –, $F(2, 3)$; ···, $F(3, 3)$. (c) —, $L(0, 1)$; —, $F(1, 1)$; – – – , $F(2, 1)$; ···,$F(3, 1)$.

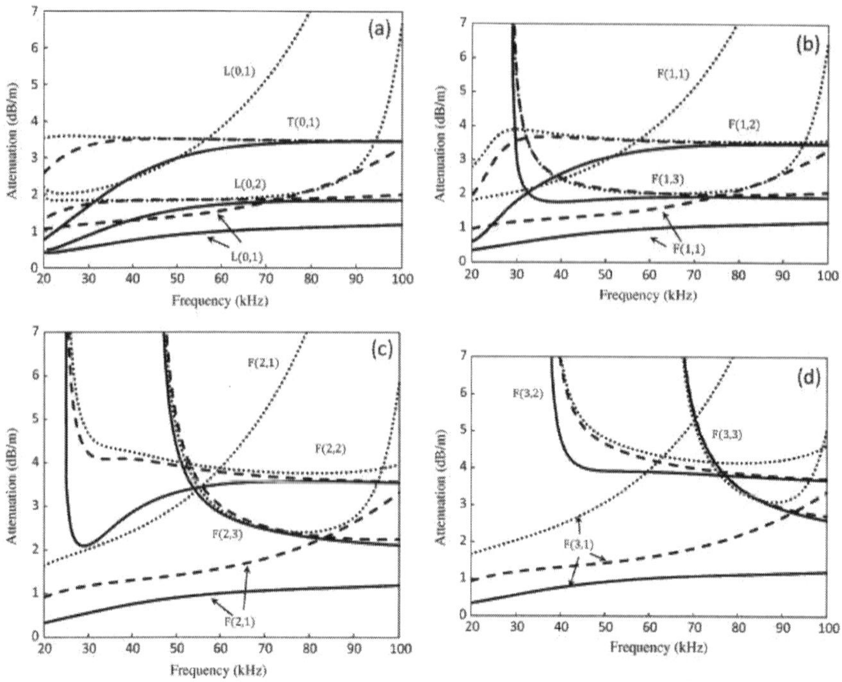

Figure. 5. Attenuation in pipe coated with viscoelastic bitumen. —, 1.5 mm coating thickness; – – –, 3 mm coating thickness; ···, 5 mm coating thickness.

4.4. Scattering from a non-axisymmetric defect

The analysis undertaken so far investigates the influence of the coating on the properties of individual modes. This includes the mode shape and wavenumber, so that if one returns to Eqs. (13) and (14) this relates to the eigenvector u_1, and the eigenvalue γ. However, this article is also interested in the effect of the coating on the way in which a wave scatters from a defect. In terms of Eqs. (13) and (14) this translates into the effect of the coating on the modal amplitudes A,B and *C*. To do this, the scattering from a non-axisymmetric defect is investigated, as this is more representative of actual defects found in the field when compared, say, to the axisymmetric defects studied by Kirby et al.[22] and [23]. To investigate the influence of the coating on the way in which a defect scatters the incident wave, scattering from a non-axisymmetric defect is studied here by comparing reflection coefficients obtained with and without a coating at a location immediately adjacent to the start of the defect. This choice of location is convenient because it avoids the additional attenuation imparted by the coating as one moves away from the defect, and so isolates the effect of the coating on the modal amplitudes. For this problem the properties of the coating and the substrate are the same as those described at the start of this section; the non-axisymmetric defect then extends 10% around the circumferential direction of the pipe, penetrates the coating and goes 50% into the pipe wall (2.8 mm), and has a length of 2.5 mm. The area ratio of the defect (cross-sectional area of the defect compared to cross-sectional area of the pipe without coating) is approximately 5%.

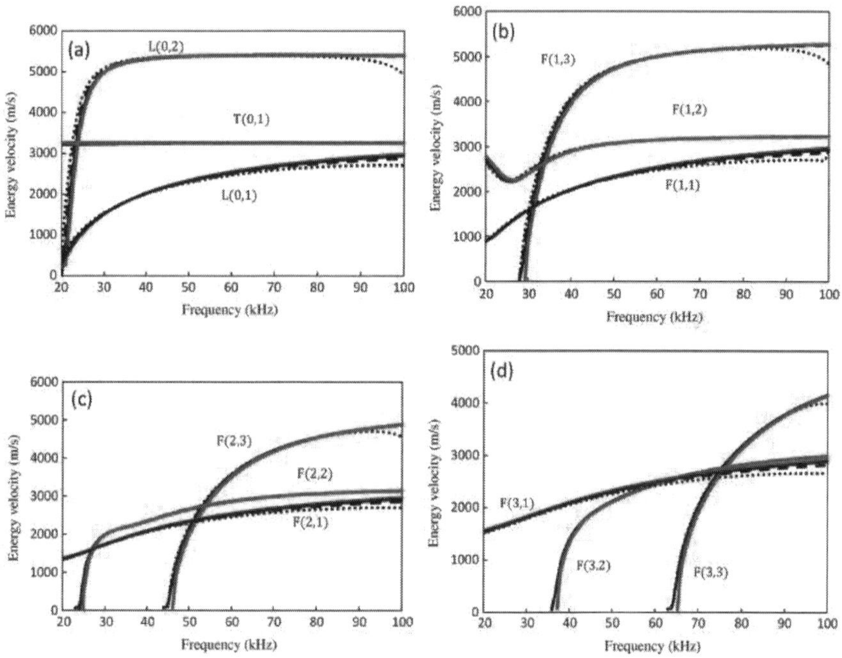

Figure. 6. Comparison between energy velocities for an uncoated and coated pipe. –, uncoated pipe; —, 1.5 mm coating thickness; – – –, 3 mm coating thickness; ···, 5 mm coating thickness.

Fig. 7(a) and (b) shows the reflection coefficients for the *T*(0, 1) and *L*(0, 2) family of modes with excitation from T(0,1) and L(0,2) modes, respectively. The reflection coefficients of flexural modes are calculated using the maximum displacement values around the circumference of the pipe. In Fig. 8(a) and (b) a similar comparison is also made but this time for a different defect, which has the same length and depth as the defect in Fig. 7, but in Fig. 8 the defect extends around 50% of the pipe circumference. The area ratio of the defect shown in Fig. 8 is therefore 25%. It is clear in both figures that the reflection coefficients with and without a coating are very similar, and this means that the modal amplitudes are very similar as well. Thus, the coating does not affect the way in which the wave is scattered by the defect, apart from the additional change in geometry caused by the thickness of the coating itself. That is, there is no additional mode conversion taking place because of the presence of the coating, and the influence of the coating is restricted to the attenuation of energy as the wave propagates down the pipe. This is an important result as it means that when undertaking experimental measurements one need only account for the effect of the coating on modal attenuation, without the need to consider additional complexities such a change in mode conversion and/or energy velocity. Thus, one does not need to model a three dimensional coated pipe if the coating is relatively thin, instead it is necessary only to model an uncoated pipe and simply to factor in modal attenuation after undertaking a SAFE solution for a coated pipe.

It is also observed in Fig. 7 and Fig. 8 that the reflection coefficient for flexural modes is stronger than for the equivalent *T*(0, 1) or *L*(0, 2) mode. However, it should be remembered that flexural modes are normally more dispersive than axisymmetric modes (see Fig. 5 and Fig. 6) and so in the time domain their modal amplitude will be reduced by dispersion as they propagate along the pipe, which means that the reflection coefficient for flexural modes is likely to be significantly reduced at locations well away from the defect. Note also that the sharp spikes seen in Fig. 7 and Fig. 8 are caused by modal cut-on in the uncoated pipe; the additional damping provided by the coating is seen to reduce the amplitude of these peaks and in experiments one may expect to see this amplitude reduced still further by additional damping in the steel substrate that is not accounted for in this model. Fig. 7 and Fig. 8 also illustrate how a change in shape of this type of defect significantly alters the energy distribution among flexural modes.

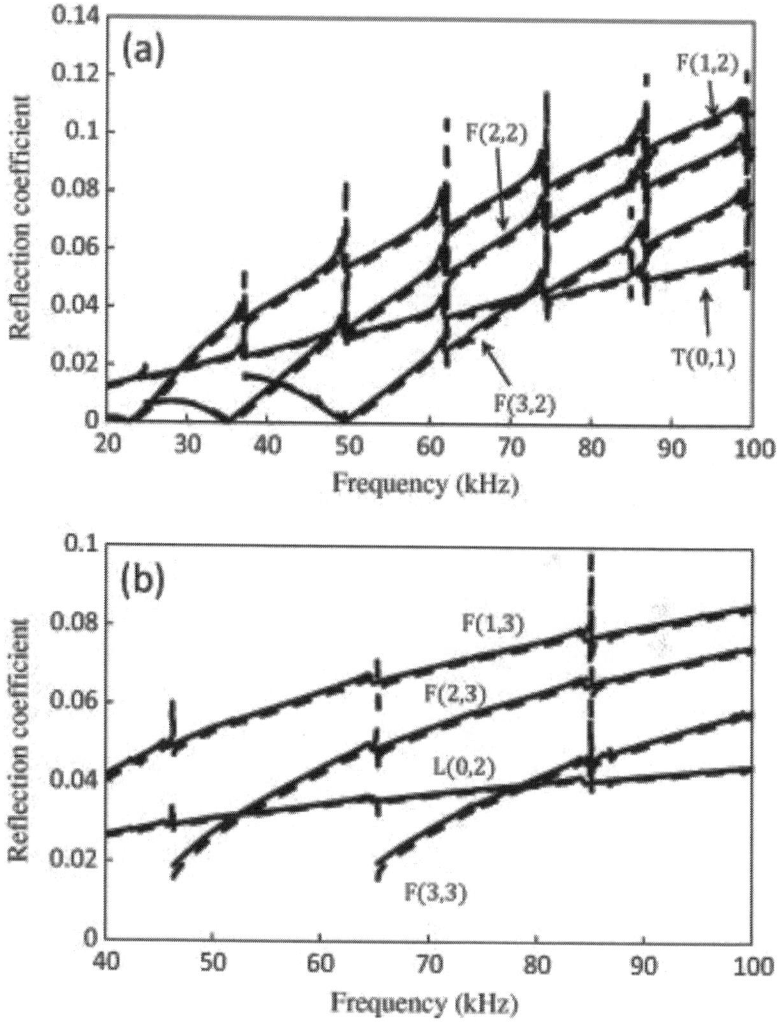

Figure. 7. Reflection coefficients for (a) the $T(0, 1)$ mode and (b) the $L(0, 2)$ mode incident upon a 5% area ratio non-axisymmetric defect: —, coated pipe; – – –, uncoated pipe.

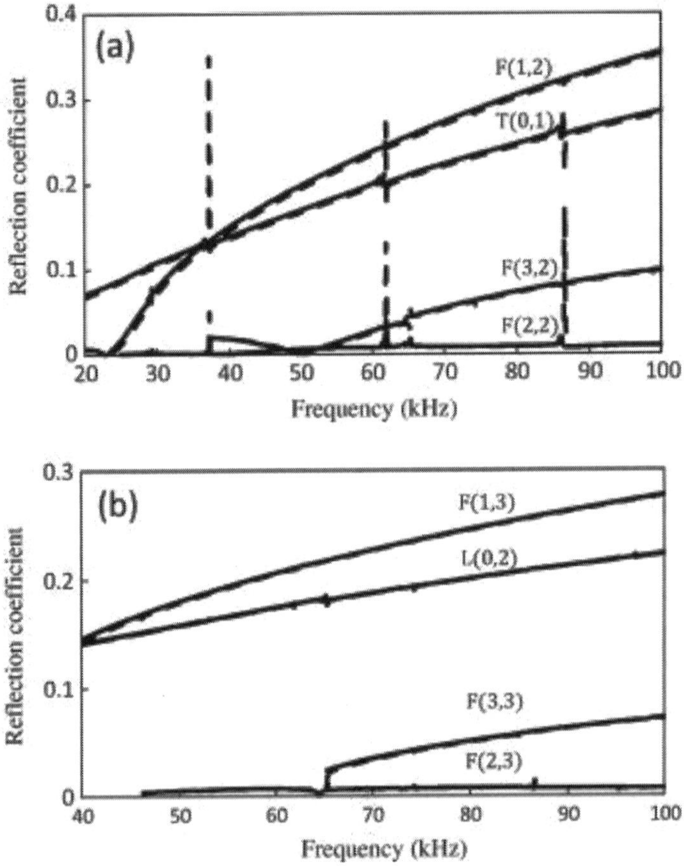

Figure. 8. Reflection coefficients for (a) the *T*(0, 1) mode and (b) the *L*(0, 2) mode incident upon a 25% area ratio non-axisymmetric defect: —, coated pipe; – – –, uncoated pipe.

5. TIME DOMAIN: RESULTS AND DISCUSSION

The previous section investigated wave propagation in the frequency domain; however, LRUT is normally undertaken in the time domain because this is a convenient way to analyse scattering from a defect. Operating in the time domain does, however, mean that the user is faced with the task of interpreting a complex signal and understanding these signals is often a significant challenge. To investigate the interpretation of these signals it is important to generate predictions in the time domain, and here we will follow the method of Duan and Kirby [21]. Moreover, time domain predictions have yet to appear for coated pipes and so in view of the practical importance of time domain signatures this article also shows how time domain predictions may be generated and reviews the way in which the coating is likely to influence the interpretation of time domain data.

Time domain predictions are generated here using an incident pulse consisting of a 10 cycle Hanning windowed sinusoidal wave with a centre frequency of 70 kHz. This pulse is then transformed into the frequency domain using a discretised Fourier transform with 1836 discrete frequencies extending from 35 kHz to 105 kHz in increments of 38.15 Hz. The complex amplitude obtained for the pulse at each frequency is then used as the incident modal amplitude in Eq. (15). Following the solution of Eq. (15) at each discrete frequency, an inverse Fourier transform is used to generate the time domain predictions. The number of discrete frequencies chosen, as well as the size of these frequency increments, are designed to minimise numerical noise whilst at the same time maintaining an acceptable solution time.

The sample problem investigated in this section is based on the non-axisymmetric scattering problem studied in the previous section. Thus, the defect extends around 20% of the circumference and penetrates through the steel substrate and has a length of 2.5 mm. All other parameters for the pipe and the coating remain the same as those detailed at the start of Section 3.2. Analysing scattering from a defect in the time domain is complicated by modes travelling at similar energy velocities so that they overlap one another. Therefore, in order to separate out modes in the figures that follow, the scattered signal is calculated at a distance of 5 m from the defect and at two separate circumferential locations, x=R,y=0 and x=0,y=R, where R is the outer radius of the uncoated pipe. Note that y=R is directly opposite to the defect, see Fig. 1. These two circumferential locations are also referred to as '*x* direction' and '*y* direction' in Figs. 6 and 7 of Ref. [21]. Accordingly, in Fig. 9 the predicted displacement at x=R and y=R is shown for scattering by the non-axisymmetric defect in a coated and uncoated pipe at a distance of 5 m from the defect. The defect is excited using *T*(0, 1) and so only the circumferential displacement is shown in Fig. 9. In this figure, the amplitude of the reflected pulse is seen to significantly reduce following the addition of the coating, which is what one would expect to see following the analysis in the previous section. Of course, the size of this reduction is largely dictated by the effective distance between source and receiver, which in this case is 10 m. However, it is noted that the results in the previous section indicate that no additional mode conversion takes place because of the presence of the coating and so one may infer that the drop in amplitude seen in Fig. 9 is solely caused by modal attenuation. For example, in this problem the attenuation of *T*(0, 1) at 70 kHz is 3.44 dB/m, which gives a sound power reduction of 34.4 dB over 10 m. If one reads the attenuation directly from Fig. 9 then the value obtained for *T*(0, 1) is 32.3 dB. Thus the two figures are close enough to one another to support previous observations regarding the influence of the coating on mode conversion, with the small difference seen here likely to be caused by truncation and discretisation errors in the discrete Fourier transform.

It is well known that problems are found when attempting to detect defects in coated pipes, and these problems are clearly evident in Fig. 9. Whilst the dispersive behaviour of each mode changes very little after adding the coating, at least within this frequency range, the amplitude of each mode reduces significantly and this is why the use of *T*(0, 1) on coated pipes is likely to be very difficult. However, it was shown in the previous section that *L*(0, 2) may be more useful and so in Fig. 10 the equivalent response of a pipe excited by *L*(0, 2) is shown. Again, the amplitude of each mode is seen to reduce following the addition of the coating, although it is noticeable that this also means that the noisy "tail" is also damped down. Clearly, the reflected amplitude of the *L*(0, 2) family is higher than that seen for *T*(0, 1), and here the attenuation of *L*(0, 2) is now 17.2 dB when compared to 32.3 dB for *T*(0, 1). This difference in amplitude is significant in the context of real experimental measurements, although the figures seen here are of course relevant only to this type of problem. Obviously, if one changes the type of defect, the coating and/or the distance between the defect and the receiver then these figures will change, however the results obtained in the previous section illustrate that the properties of the $L(0, 2)$ mode should normally deliver lower levels of attenuation, at least over the frequency range typically associated with LRUT.

The time domain signatures shown in Fig. 9 and Fig. 10 illustrate that, in principle, there should be little difference in the complexity of a pulse reflected from a defect in a coated or uncoated pipe. However, this is not normally the experience in real experiments taken on pipes that have been *in situ* for a number of years. It is likely, therefore, that additional levels of complexity are caused by an increase in noise that arises from the coating itself. For example, it is possible that over time most coatings in real applications will start to degrade, so that small inhomogeneities appear, as well as delamination and/or de-bonding from the pipe. These imperfections will impart additional levels of attenuation, as well as introduce additional scattering. Thus, it is these imperfections that are likely to be the cause of additional problems with LRUT in coated pipes and not issues such as mode conversion or additional dispersion. Accordingly, it remains to be seen how the suggested improvements identified in this article are affected by imperfections within the coating of real pipe systems.

6. CONCLUSIONS

This article has used a hybrid SAFE-finite element based method to analyse the scattering of guided waves from non-axisymmetric defects in coated pipes. It is shown that predictions in the time and frequency domain can be generated for an infinite pipe using relatively modest computational facilities. The predictions obtained from this model indicate that the torsional [$T(0, 1)$] family of modes have high levels of attenuation for the frequency range normally used in LRUT (20–120 kHz); moreover, the levels of attenuation for torsional modes are very sensitive to the thickness of the coating in the low frequency region. The longitudinal family of modes [$L(0, 2)$] are seen to have lower levels of attenuation and so it is observed that for the viscoelastic coating studied here the longitudinal family of modes will have a longer range when compared to torsional modes. It is also observed that the energy velocity for each mode is very similar to the group velocity of the equivalent mode for an uncoated pipe. This means that the coating does not introduce additional levels of dispersion beyond what is normally observed for uncoated pipes, at least over the typical frequency range used in LRUT. This is an important result when attempting to interpret signals obtained in the time domain.

The frequency and time domain predictions generated for a non-axisymmetric defect demonstrate that scattering from the defect remains largely unaffected by the presence of the coating. In other words, there is no additional mode conversion occurring in the pipe because of the presence of the coating, and the only effect a coating has on scattering is caused by the small change in the geometry of the defect, which may be neglected for most coatings. Thus, it appears reasonable to conclude that the only significant effect that needs to be considered is the attenuation the coating imparts on each mode. Therefore, if one wishes to model the effect of a viscoelastic material on scattering from a defect in a pipe or other structure then provided the thickness of the material is small compared to the thickness of the substrate, it is sufficient to model only the uncoated structure and to account for the influence of the viscoelastic material through modal attenuation, which may readily be found using the SAFE method. Furthermore, when undertaking experimental measurements the methods currently used in LRUT may also be retained, such as the interpretation of the returning signal using group velocity data, and it is necessary only to correct for the attenuation of each mode when reviewing echoes from a defect in the time domain.

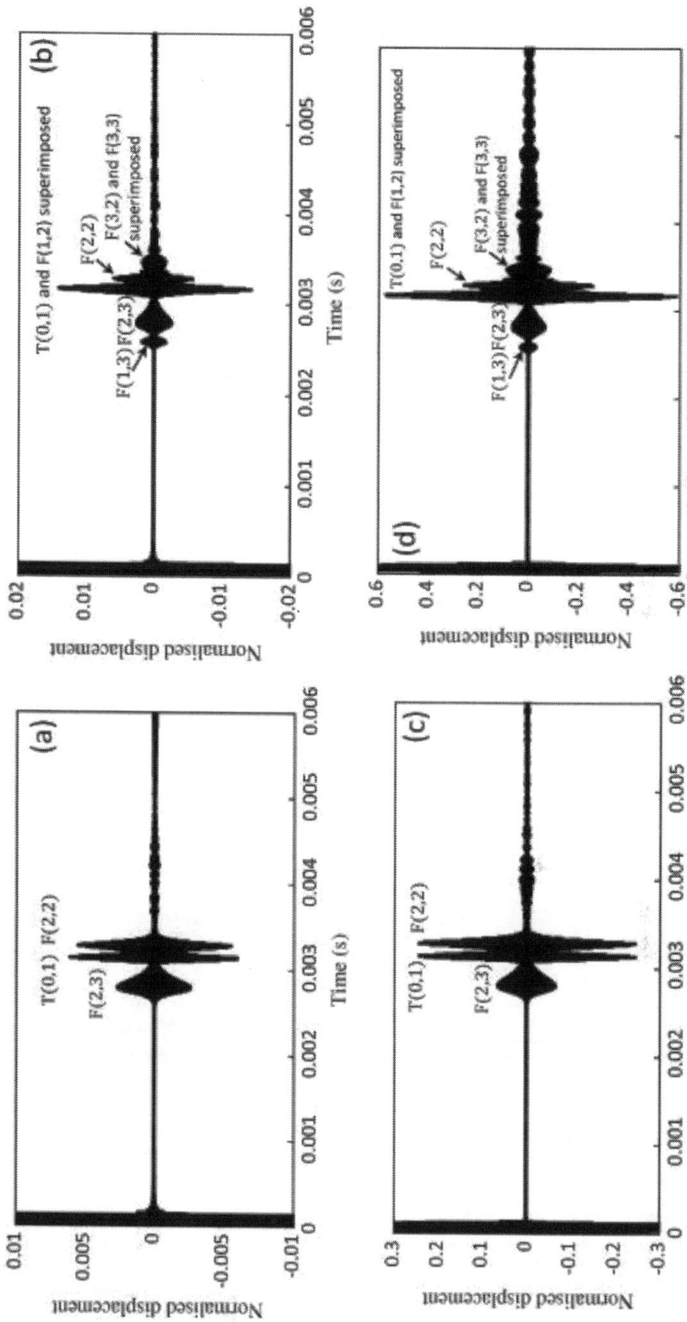

Figure. 9. Predicted circumferential displacement for the *T*(0, 1) mode incident upon a non-axisymmetric defect. (a)x=R; (b) y=R, coated; (c) x=R; (d) y=R, uncoated.

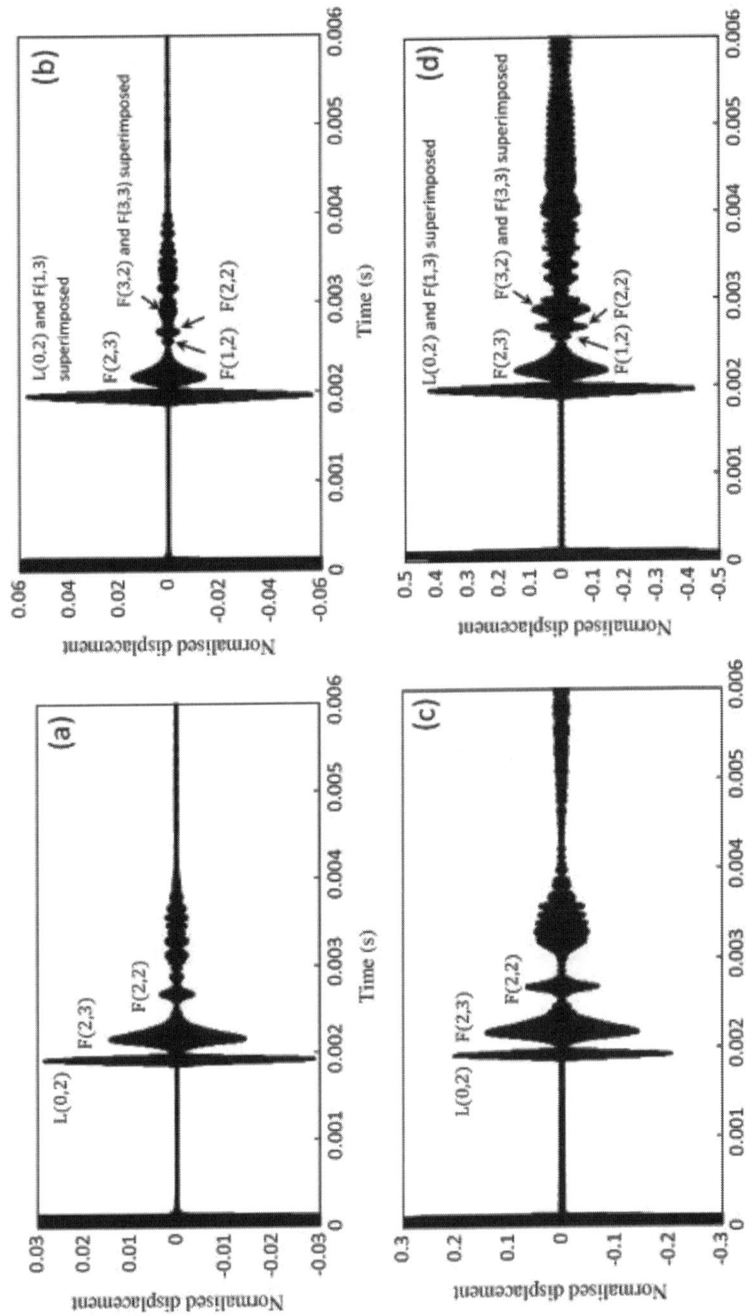

Figure. 10. Predicted axial displacement for the *L*(0, 2) mode incident upon a non-axisymmetric defect. (a) x=R; (b)y=R, coated; (c) x=R; (d) y=R, uncoated.

The theoretical method developed here is designed to take advantage of the long slender structures found in pipelines. This has enabled numerical predictions to be generated efficiently in the frequency and the time domain using relatively modest computer facilities. However, in the future it may be interesting to investigate whether the computational efficiency of this method may be increased through the introduction of higher order elements such as those described by Ostachowicz el al. [17] and/or through additional optimisation of the finite element mesh.

REFERENCES

1. J.J. Ditri, J.L. Rose, Excitation of guided elastic wave modes in hollow cylinders by applied surface tractions, J. Appl. Phys. 72 (7) (1992) 2589–2597.

2. M.J.S. Lowe, D.N. Alleyne, P. Cawley, Defect detection in pipes using guided waves, Ultrasonics 36 (1998) 147–154.

3. N. Barshinger, J.L. Rose, Guided wave propagation in an elastic hollow cylinder coated with a viscoelastic material, IEEE Trans. Ultrason. Ferroelectr. Freq. Control 51 (2004) 1547–1556.

4. I. Bartoli, A. Marzani, F. Lanza di Scalea, E. Viola, Modeling wave propagation in damped waveguides of arbitrary cross-section, J. Sound Vib. 295 (2006) 685– 707.

5. T. Hayashi, W.-J. Song, J.L. Rose, Guided wave dispersion curves for a bar with an arbitrary cross-section, a rod and rail example, Ultrasonics 41 (2003) 175– 183.

6. S. Finnveden, Evaluation of modal density and group velocity by a finite element method, J. Sound Vib. 273 (2004) 51–75.

7. V. Damljanovic′, R.L. Weaver, Propagating and evanescent elastic waves in cylindrical waveguides of arbitrary cross section, J. Acoust. Soc. Am. 115 (2004) 1572–1581.

8. F. Treyssède, Elastic waves in helical waveguides, Wave Motion 121 (2008) 457–470.

9. F. Treyssède, L. Laguerre, Numerical and analytical calculation of modal excitability for elastic wave generation in lossy waveguides, J. Acoust. Soc. Am. 133 (2013) 3827–3837.

10. J. Mu, J.L. Rose, Guided wave propagation and mode differentiation in hollow cylinders with viscoelastic coatings, J. Acoust. Soc. Am. 124 (2008) 866–874.

11. M. Castaings, M. Lowe, Finite element model for waves guided along solid systems of arbitrary section coupled to infinite solid media, J. Acoust. Soc. Am. 123 (2008) 696–708.

12. A. Marzani, E. Viola, I. Bartoli, F. Lanza di Scalea, P. Rizzo, A semi-analytical finite element formulation for modelling stress wave propagation in axisymmetric damped waveguides, J. Sound Vib. 318 (2008) 488–505.

13. A. Bernard, M.J.S. Lowe, M. Deschamps, Guided waves energy velocity in absorbing and non-absorbing plates, J. Acoust. Soc. Am. 110 (2001) 186– 196.

14. J. Hua, J.L. Rose, Guided wave inspection penetration power in viscoelastic coated pipes, Insight 52 (2010) 195–200.

15. M.V. Predoi, M. Castaings, L. Moreau, Influence of material viscoelasticity on the scattering of guided waves by defects, J. Acoust. Soc. Am. 124 (5) (2008) 2883–2894.

16. A. _ Zak, A novel formulation of a spectral plate element for wave propagation in isotopic structures, Finite Elem. Anal. Des. 45 (2009) 650–658.

17. W. Ostachowicz, P. Kudela, M. Krawczuk, A. _ Zak, Guided Waves in Structures for SHM, John Wiley & Sons Ltd., U.K, 2012 (Chapter 2).

18. J.M. Galán, R. Abascal, Remote characterization of defects in plates with viscoelastic coatings using guided waves, Ultrasonics 42 (2004) 877– 882.

19. W.J. Zhou, M.N. Ichchou, J.M. Mencik, Analysis of wave propagation in cylindrical pipes with local inhomogeneities, J. Sound Vib. 319 (2009) 335– 354.

20. F. Benmeddour, F. Treyssède, L. Laguerre, Numerical modelling of guided wave interaction with non-axisymmetric cracks in elastic cylinders, Int. J. Solids Struct. 48 (2011) 764–774.

21. W. Duan, R. Kirby, A numerical model for the scattering of elastic waves from a non-axisymmetric defect in a pipe, Finite Elem. Anal. Des. 100 (2015) 28–40.

22. R. Kirby, Z. Zlatev, P. Mudge, On the scattering of torsional elastic waves from axisymmetric defects in coated pipes, J. Sound Vib. 331 (2012) 3989– 4004.

23. R. Kirby, Z. Zlatev, P. Mudge, On the scattering of longitudinal elastic waves from axisymmetric defects in coated pipes, J. Sound Vib. 332 (2013) 5040– 5058.

24. B.A. Auld, Acoustic Fields in Waves and Solids, vol. 1, Krieger, Malabar, FL, 1990.

CHAPTER 4

Propagation of Ultrasonic Guided Waves in Composite Multi-Wire Ropes

Renaldas Raisutis *, Rymantas Kazys, Liudas Mazeika, Vykintas Samaitis and Egidijus Zukauskas

Ultrasound Institute of Kaunas University of Technology, K. Barsausko 59, Kaunas Lt-51423, Lithuania

ABSTRACT

Multi-wire ropes are widely used as load-carrying constructional elements in bridges, cranes, elevators, *etc*. Structural integrity of such ropes can be inspected by using non-destructive ultrasonic techniques. The objective of this work was to investigate propagation of ultrasonic guided waves (UGW) along composite multi-wire ropes in the cases of various types of acoustic contacts between neighboring wires and the plastic core. The modes of UGW propagating along the multi-wire ropes were identified using modelling, the dispersion curves were calculated using analytical and semi-analytical finite element (SAFE) techniques. In order to investigate the effects of UGW propagation, the two types of the acoustic contact between neighboring wires were simulated using the 3D finite element method (FE) as well. The key question of investigation was estimation of the actual boundary conditions between neighboring wires (solid or slip) and the real depth of penetration of UGW into the overall cross-section of the rope. Therefore, in order to verify the results of FE modelling, the guided wave penetration into strands of multi-wire rope was investigated experimentally. The performed modelling and experimental investigation enabled us to select optimal parameters of UGW to be used for non-destructive testing.

KEYWORDS

multi-wire ropes; modelling; ultrasonic guided waves; dispersion curves

1. INTRODUCTION

Multi-wire ropes are widely used in different sectors of industry as load-carrying constructional elements in bridges, cranes, elevators, *etc.* Due to environmental and operational conditions and fatigue caused by rotational stresses, multiple defects may develop in the internal structure of multi-wire ropes. One of the promising methods for investigation of non-homogeneities of the internal structure of wire ropes can be application of ultrasonic guided waves (UGW) [1,2,3,4,5,6,7,8,9,10]. UGW are suitable for detection of various structural non-homogeneities, because they can propagate long distances and are quite sensitive both to surface and internal defects [11,12,13]. However, in the case of multi-wire ropes, the key issues are whether UGW will propagate only in one particular wire/strand, in which it had been excited, or in the whole composite multi-wire rope, and how the penetration of UGW will depend on the excitation method and the type of a mechanical contact between neighboring wires.

The objective of this work was to investigate the propagation of ultrasonic guided waves (UGW) along composite multi-wire ropes in the cases of various types of acoustic contacts between neighboring wires and the plastic core.

2. THE OBJECT OF INVESTIGATION

A multi-wire steel rope to be investigated using ultrasound by itself is a complex composite structure in which multiple reflections, scattering, and the presence of multiple wave modes of UGW occur. Therefore, in order to understand the effects of UGW propagating along the multi-wire rope and penetration into the internal structure, determination of type of the acoustic contact between neighboring wires is necessary. In addition, the optimal parameters of UGW used for non-destructive testing of the particular objects are necessary to be investigated and selected.

As an object of investigation, a multi-wire steel rope consisting of six strands, each with the diameter of 11.15 mm, had been selected. The core of the multi-wire rope had been filled with polypropylene, and each strand consisted of a bundle of approximately 31 wires, each with the diameter of 2.05 mm. The cross-sectional view of the investigated composite multi-wire rope is presented in Figure 1. In order to simplify further analysis the assumption was made that the structure of rope consists of straight strands.

2.1. Dispersion Curves of Propagating Guided Waves

Dispersion curves of propagating UGW along the object under investigation show the distribution of phase or group velocities in a particular frequency bandwidth. Therefore, in order to identify the particular modes of UGW, which propagate along the object under investigation and select suitable ones for non-destructive testing, the dispersion curves of phase and group velocities must be calculated. There are several techniques for calculation of the dispersion curves of UGW: semi-analytical finite element (SAFE) technique [14,15,16,17,18,19,20] and analytical ones [21]. The SAFE technique is more attractive for analysis of the objects with an arbitrary geometry of the cross-section (e.g., rails, bars, *etc.*). The SAFE technique has been used by Treyssede and Laguerre for investigation of UGW propagation along multi-wire helical strands with a metal cylindrical core [4,7,9,10,14]. The wires were assumed to be solidly connected [4,9,14]. They have found that the wave numbers and energy velocities of axisymmetric modes, such as $L(0,1)$ and $T(0,1)$, also flexural mode $F(1,1)$ at low frequencies and small helix lay angles (less than 15 degrees) of helical waveguides are very similar to the straight (not twisted) and twisted (helical) wires of multi-wire helical waveguides [4,9,14]. However, the cut-off region of the $L(0,1)$-like mode was observed at a notch frequency [4,9,14]. The advantage of the analytical technique, comparing it to SAFE, is a possibility to determine the frequency dependent guided wave leakage losses in the case of submerged wire ropes, for instance, as in cement grout or rock-embedded wire ropes [20,21].

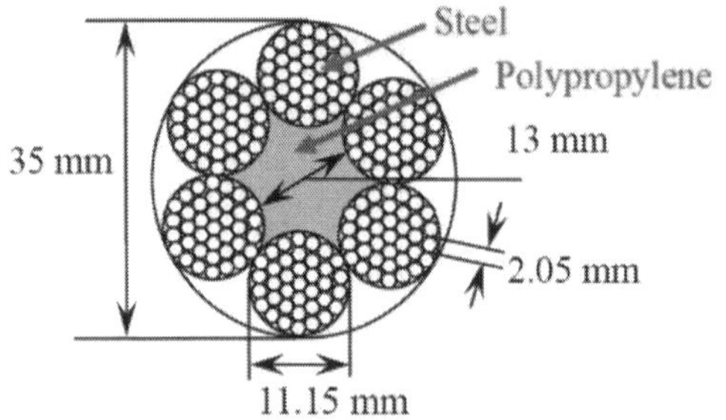

(a)

(b)

Figure 1. The rope under investigation: (**a**) schematic view; and (**b**) photo of the cross-section of the multi-bundle and multi-wire twisted rope.

In order to calculate dispersion curves for a multi-wire rope with a polymer core, several tasks are necessary to be solved. At first, the stranding of wires was eliminated and calculation of dispersion curves was divided into several stages. During the first stage, the dispersion curves of UGW were calculated using the SAFE technique for a single wire having the diameter of 2.05 mm (Figure 2a,b). The material properties used for the SAFE calculations of the dispersion curves are summarized in Table 1.

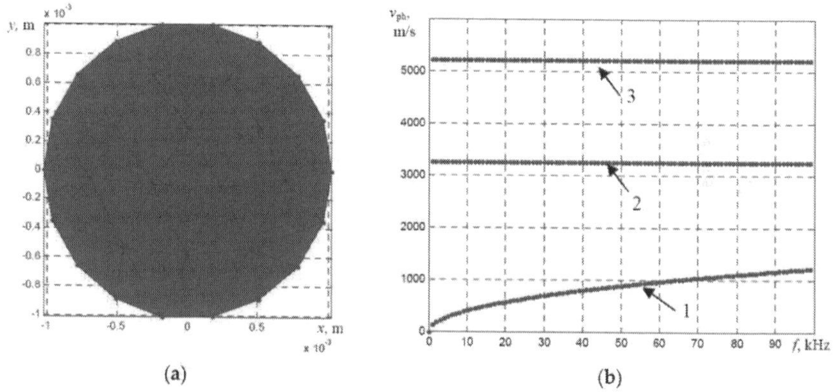

Figure 2. The mesh of the SAFE model of a single wire with diameter of 2.05 mm (**a**) and the dispersion curves of the phase velocity of guided wave modes (**b**): 1—flexural*F*(1,1) mode; 2—torsional *T*(0,1) mode; 3—longitudinal *L*(0,1) mode.

Table 1. Material properties between neighboring wires used for the calculation of the dispersion curves.

Material	Young's Modulus	Poisson's Ratio	Density
Steel	212.8 GPa	0.287	7800 kg/m^3
Polypropylene core	3.54 GPa	0.36	901 kg/m^3
"Elastic fluid"	0.075 MPa	0.499	1000 kg/m^3

During the SAFE calculations the triangular finite elements were used in order to obtain the desired accuracy, the average spatial size of element was set to 1 mm, which corresponds to 15 elements per wavelength, according to the minimum wavelength of the slowest $F(1,1)$ mode at 100 kHz frequency.

Calculation results of UGW propagation along a single wire having the diameter of 2.05 mm (Figure 2a) demonstrated that only fundamental modes propagate in the frequency band up to 100 kHz (Figure 2b). The dispersion curves also show that, in the selected frequency region, the fastest mode is longitudinal $L(0,1)$ and the slowest mode—flexural $F(1,1)$, whereas the torsional mode $T(0,1)$ is not dispersive.

The obtained results demonstrated that in this frequency range only fundamental modes propagate in a single wire. Typically, a multi-wire rope structure is more complicated, so, during the second stage the dispersion curves of the strand with 31 wires (the diameter of each wire 2.05 mm) having the overall diameter of 11.15 mm were calculated. In this case it was assumed that there is a solid acoustic contact between neighboring wires at their contact points. The cross-section of the meshed strand of 31 wires (the overall diameter 11.15 mm) is presented in Figure 3a. The obtained phase velocity dispersion curves and the comparison with the dispersion curves of the single wire (diameter 2.05 mm) are presented in Figure 3b. It should be noted that in the case of a bundle of wires or multi-wire rope the modes should be named as the axial-like $L(0,1)$, the flexural-like $F(1,1)$, and the torsional-like $T(0,1)$ modes.

However, the numerical analysis of the multi-wire rope with the six strands consisting of 31 wires each and the polymer core is quite complicated due to limited computational resources of SAFE and the expected memory overflow. Therefore, the assumption was made that each strand can be replaced by a solid rod, having the same diameter (11.15 mm). The comparison of the dispersion curves of the UGW propagating in a multi-wire strand of 31 wires and in the solid rod are presented in Figure 4a,b. Comparison of these two cases demonstrate that the phase velocities of the *F*(1,1) and *L*(0,1) modes are similar and the velocity of the *T*(0,1) mode in the strand of 31 wires is reduced by 23% in comparison to the phase velocities in the case of a solid rod. The calculation results demonstrate that a solid circular rod having the diameter of 11.15 mm is not fully equivalent to the meshed bundle of 31 wires but this replacement can be done for analysis of fundamental modes of UGW (especially *L*(0,1) and *F*(1,1)) at frequencies below 100 kHz.

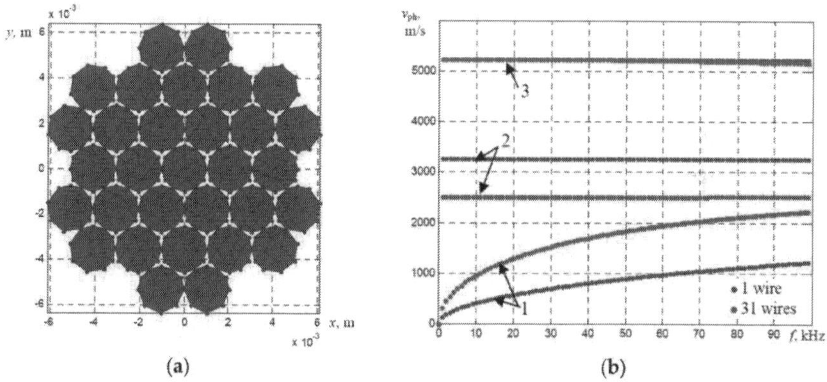

Figure 3. The mesh of the SAFE model of a bundle of solidly connected 31 wires (overall diameter 11.15 mm) (**a**) and comparison of the phase velocity dispersion curves of the guided wave modes (**b**) propagating along: one wire (**blue**), bundle of 31 wires (**red**): 1—flexural-like *F*(1,1) mode; 2—torsional-like *T*(0,1) mode; 3—axial-like*L*(0,1) mode.

This comparative analysis was made in order to be sure that replacement of the bundle of solidly-connected 31 wires by a single solid rod gives the same or similar results of the calculated dispersion curves. Then the dispersion curves of the whole multi-wire rope could be calculated as the curves of a bundle of solidly-connected six strands (the overall diameter 35 mm). This simplified assumption could reduce the required computational resources and additionally enable us to estimate frequency dependent leakage losses using the simplified analytical technique.

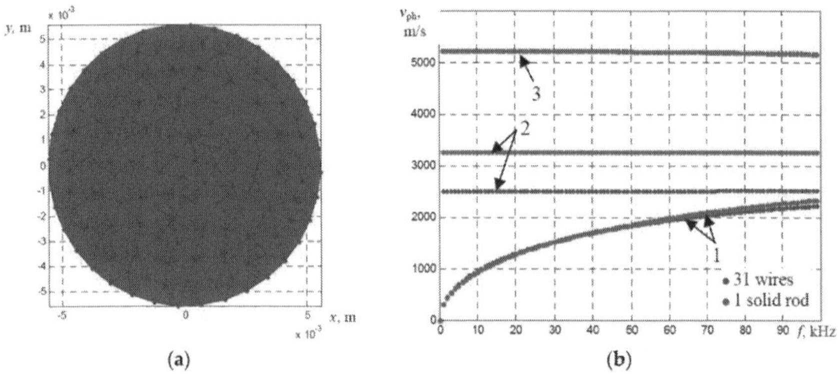

Figure 4. The mesh of the SAFE model of a strand that assumed to be equivalent to the bundle of solidly connected 31 wires having the overall diameter of 11.15 mm (**a**) and comparison of the phase velocity dispersion curves of the guided wave modes (**b**) propagating along the bundle of 31 wires (**blue**) and the solid rod of same diameter (**red**): 1—flexural-like $F(1,1)$ mode; 2—torsional-like $T(0,1)$ mode; 3—axial-like $L(0,1)$ mode.

When conducting theoretical analysis on a multi-wire steel rope of a complex structure, an important question arises which is related to the acoustic contact between neighboring wires. Transmission of the appropriate modes between neighboring wires depends on a type of the acoustic contact. For example, in the case of a slip contact there is no transmission of the axial component of the particle velocity of the propagating mode to the neighboring wires. Therefore, only the component of the particle velocity perpendicular to the axis of the rope could be transmitted to the neighboring wires. In the case of a solid contact, the axial component is transmitted to neighboring wires. In order to investigate the influence of the acoustic contact on dispersion curves of the guided waves, the calculations for the simplified structure of the wire rope with the polymer core was performed. The calculations of the dispersion curves were performed for two separate cases. In Case A, it was assumed that the acoustic contact between neighboring wires is solid and in Case B a thin layer of a new material named as "elastic fluid" (Table 1) with the shear wave velocity close to zero was introduced between the separate strands to simulate the slip contact. The cross-section of the meshed multi-wire rope in the cases of a solid and slip contacts are presented in Figure 5a,b. The dispersion curves in the case of a solid and slip contact are presented in Figure 5c.

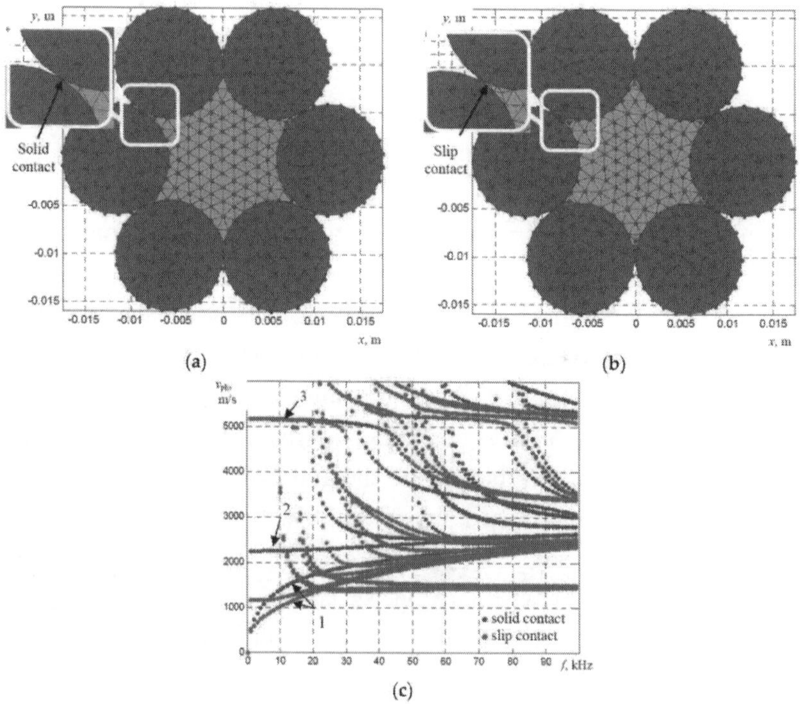

Figure 5. The mesh of the SAFE model of a multi-wire rope in the case of a solid (**a**) and a slip (**b**) contact between neighboring strands and comparison of the phase velocity dispersion curves of the guided wave modes (**c**) propagating along the particular structure: solid contact (**blue**), slip contact (**red**): 1—flexural-like *F*(1,1) mode; 2—torsional-like *T*(0,1) mode; 3—axial-like *L*(0,1) mode.

The results obtained demonstrate that the type of the contact between neighboring straight strands influences the type of modes that propagate in the structure. The phase velocity of the $L(0,1)$-like mode is the same for both cases at the frequencies below 30 kHz, whereas the phase velocity of the asymmetrical $F(1,1)$-like mode is shifted down in the case of the slip contact. It can be seen that in the case of the slip contact a large number of higher-order modes propagate in the structure at the frequencies above 10 kHz.

2.2. Investigation of the Acoustic Contact Between Wires Using 3D FE Modelling

Dispersion curves of UGW show what modes may propagate along a multi-wire rope. The most appropriate way to excite UGW in such structures is to apply a normal (perpendicular to the axis of the rope) or tangential force to the outer surface of a wire rope (parallel to the axis of the rope). It is a promising solution for practical applications, based on a single side access to the surface of the rope. Thus, the next stage of investigation must be devoted to the issue whether those UGW modes can be excited from the outer surface and whether such propagating waves penetrate into the whole structure of a wire rope or just propagate only along the outer wires. In this case, a very important task is determination of the type of the acoustic contact between neighboring wires in a rope. Each wire is attached to another one by some force and in general, two types of acoustic contact between wires should be analyzed—a solid contact and a slip one. Those contact types influence what components (perpendicular to the axis of the rope or axial) of the particle velocity can propagate from one single wire to another one and, as a consequence, determine how deeply the propagating guided wave can penetrate into the structure under investigation. In order to get the answer to these questions, two models had been constructed using the finite element method with two different types of the acoustic contact between wires—a solid and a slip contact. In the case of the solid contact vertical (v_y) and axial (v_z) components of particle velocity are transmitted. Still, in the case of the slip acoustic contact only components (v_y) perpendicular to the rope axis are transmitted.

Elastic wave propagation can be described by the Navier equation of motion in the form of matrices [22]:

$$\mathbf{M\ddot{U} + C\dot{U} + KU = F} \tag{1}$$

where M is the mass matrix, C is the damping matrix, K is the stiffness matrix, F is the load vector, U is the vector of displacements, $U^{\cdot}$ is the vector of velocity, $U^{\cdot\cdot}$ is the vector of accelerations for each degree of freedom. For the undamped forced vibrations, the Equation (1) reduces to:

$$\mathbf{M\ddot{U} + KU = F} \tag{2}$$

The equation above was solved using the commercially available ANSYS implicit software and the Newmark's time integration scheme. In the case of a solid contact it was assumed that the contact layer between the adjacent wires has the properties of steel. In the case of a slip contact the layer between wires was simulated as solid, but with the shear wave velocity close to zero. In this case the shear wave is not transmitted between adjacent wires. The properties of two types of the contact layers between neighboring individual wires are presented in Table 1.

Just for simplicity not the whole structure of the strand of the multi-wire rope but only a row of five wires located on each other were analyzed (Figure 6). The diameter of a single wire in the model was 2 mm. The length of the simulated structure was 250 mm. Wires were meshed using SOLID45 elements which are used for modelling of solid structures. Each element had eight nodes with three degrees of freedom each: translation in x, y, and z directions. The average spatial size of the elements was 0.2 mm.

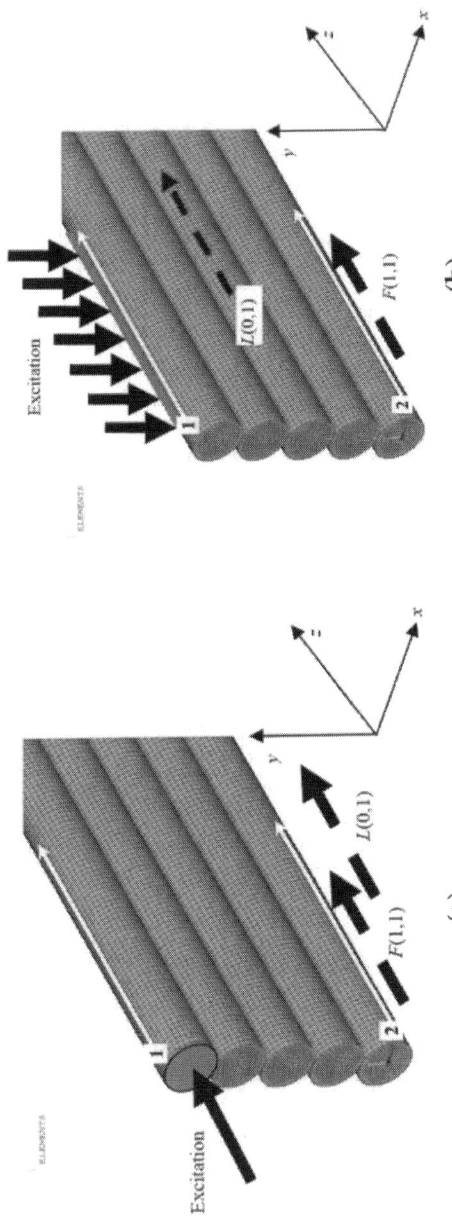

Figure 6. The sketch of FE model in the case of the edge type (symmetric) (**a**) and top (asymmetric) (**b**) excitation, when the diameter of a single wire is 2 mm and different types of the acoustic contact between neighboring wires was implemented: solid and slip. The numbers 1 and 2 denote the particular line for monitoring penetration of the guided wave into the structure.

For excitation of the UGW, the appropriate force for a three-period burst and the frequency of 60 kHz with the Gaussian envelope were applied in order to investigate wave propagation effects in more detail. Motivation of selection of 60 kHz frequency was based on the fact that, at a higher frequency, the wavelength of guided waves is two times shorter for the axial-like *L*(0,1) mode and one and a half times shorter for the flexural-like *F*(1,1) mode comparing to the case of 30 kHz. The interaction effects of propagating the *F*(1,1)-like mode with different types of the acoustic contact between the neighboring wires are stronger at the frequency of 60 kHz due to a lower ratio of displacements of radial and axial components. For the*L*(0,1)-like mode there is no significant difference due to a weak radial component. In addition, at mentioned both frequencies only the fundamental modes (e.g., the *L*(0,1)-like and the *F*(1,1)-like) propagate along the single thin wire and the influence of higher modes does not exists.

The symmetric mode in the upper wire was generated by applying a normal force to the edge of it covering the whole cross-section having the area of 3.3 mm^2 (Figure 6a). The asymmetric mode was generated applying the normal force to the top surface of the same wire, the length of the excitation region was 10 mm and the area of 10.45 mm^2 (Figure 6b). The bandwidth of the excitation pulse at −6 dB level was equal to 32 kHz. The integration step in the time domain was 0.5 µs. This step is 1/33 of the period in the case of the excitation pulse possessing the central frequency of 60 kHz.

The solution was obtained using numerical integration and in this way the particle velocity distribution in the whole simulated structure at different time instants was obtained. The results are presented in the form of B-scan images (Figure 7,Figure 8, Figure 9 and Figure 10), defining distribution of different components of a particle velocity along the central line of the upper surface of the first wire and the lower surface of the last (fifth) wire (Figure 6a,b).

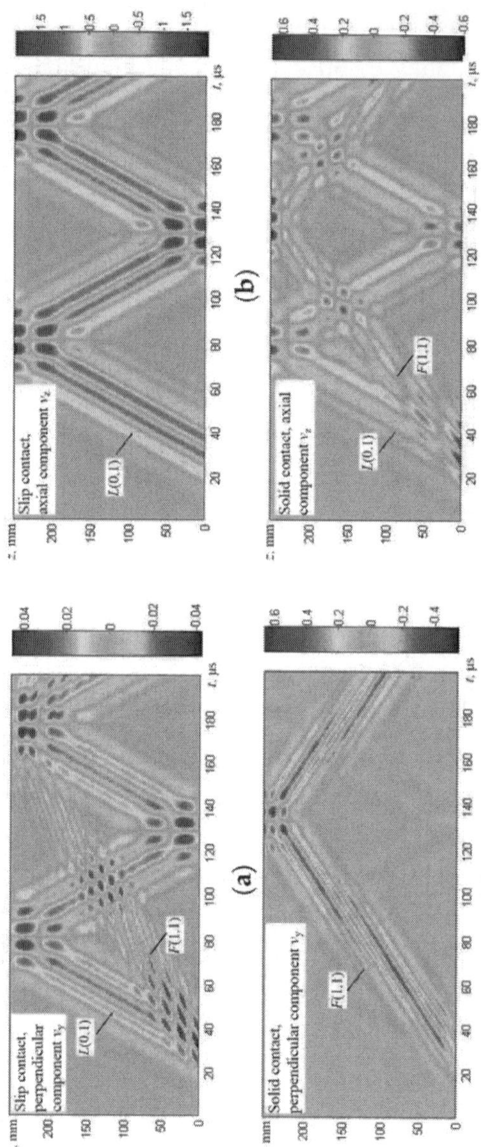

Figure 7. The B-scan images of the particle velocity along the first line of monitoring (the upper surface of the first wire) in the case of the edge type excitation: particle velocity components v_y perpendicular to the axis of the wire (**a**) and v_z (longitudinal-axial) (**b**) in the case of a slip contact and particle velocity components $v_y$perpendicular to the axis of the wire (**c**) and v_z (longitudinal-axial) (**d**) in the case of a solid contact between the neighboring wires.

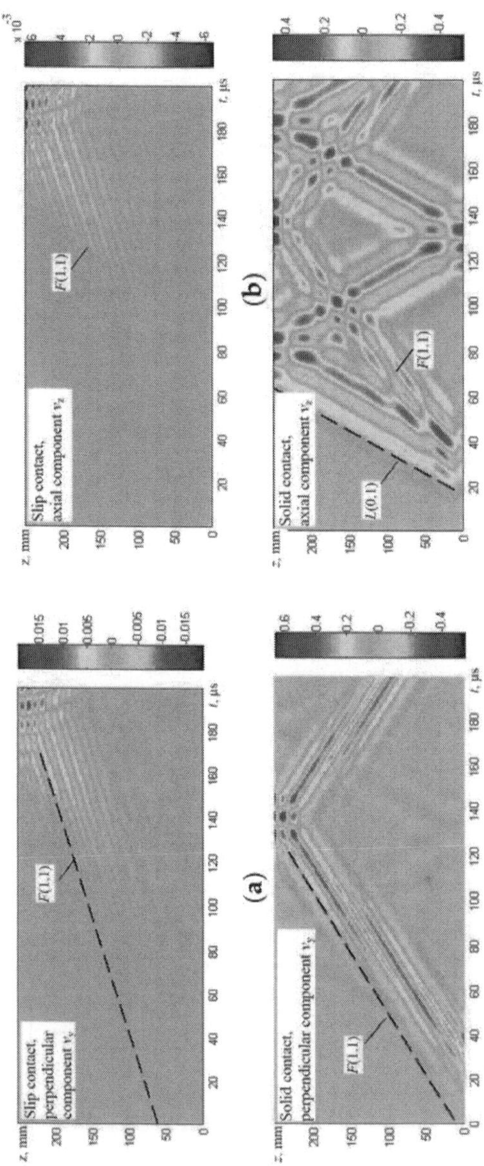

Figure 8. The B-scan images of the particle velocity along the second line of monitoring (the lower surface of the fifth wire) in the case of the edge type excitation: particle velocity components v_y perpendicular to the axis of the wire (**a**) and v_z (longitudinal-axial) (**b**) in the case of a slip contact and particle velocity components $v_y$perpendicular to the axis of the wire (**c**) and v_z (longitudinal-axial) (**d**) in the case of a solid contact between the neighboring wires.

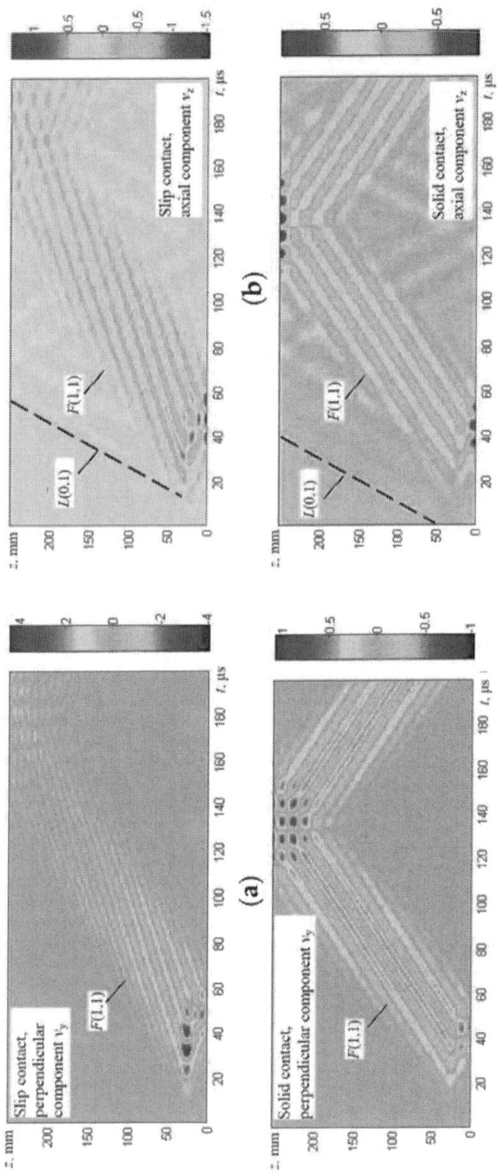

Figure 9. The B-scan images of the particle velocity along the first line of monitoring (the upper surface of the first wire) in the case of excitation from the top: particle velocity components v_y perpendicular to the axis of the wire (**a**) and v_z (longitudinal-axial) (**b**) in the case of slip contact and particle velocity components v_y perpendicular to the axis of the wire (**c**) and v_z (longitudinal-axial) (**d**) in the case of a solid contact between the neighboring wires.

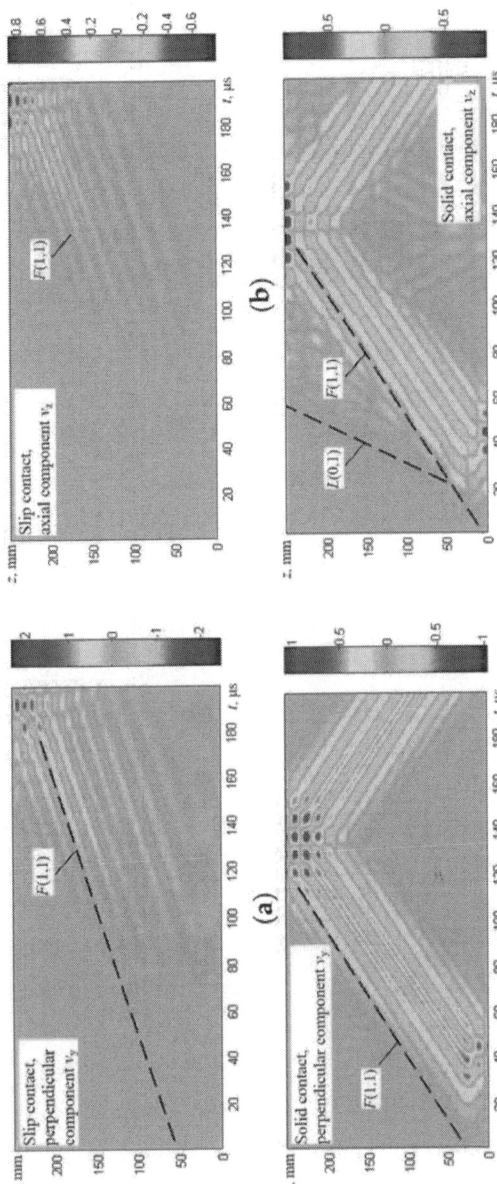

Figure 10. The B-scan images of the particle velocity along the second line of monitoring (the lower surface of the fifth wire) in the case of excitation from the top: particle velocity components v_y perpendicular to the axis of the wire (**a**) and v_z(longitudinal-axial) (**b**) in the case of the slip contact and the particle velocity components v_y perpendicular to the axis of the wire (**c**) and v_z (longitudinal-axial) (**d**) in the case of the solid contact between the neighboring wires.

It should be noted that the B-scan images have been acquired in a relatively long time interval, having a duration of 200 μs, and multiple reflections from edges of the simulated upper wire are clearly visible. The modelling results show that in the case of the edge type excitation and a slip contact between wires (Figures 7a,b and 8a,b) flexural *F*(1,1) and symmetric*L*(0,1) modes are generated in the first wire. However, only the flexural-like *F*(1,1) mode after some delay penetrates to the bottom of the structure and can be observed on the lower wires as can be seen in Figure 8a,b. In the case of a solid contact between wires and edge type excitation (Figures 7c,d and 8c,d), not only the symmetric *L*(0,1) mode, but the relatively strong asymmetric *F*(1,1) mode is generated as well, and both of them penetrate the whole structure up to the last (fifth) wire (Figure 6a).

The results obtained in the case of the top side excitation by a normal force (Figure 6b) in general demonstrate that, in both cases, the dominant flexural-like *F*(1,1) mode was generated which penetrates the whole structure (Figure 9 and Figure 10). In the case of a solid contact, the symmetric *L*(0,1) mode is simultaneously generated also.

On the basis of the simulation results, it has been found that the phase velocities of the *F*(1,1)-like mode in the cases of solid and slip contacts are different. The analysis shows that in the case of the solid contact between wires, the phase velocity of the *F*(1,1)-like mode is around 1800 m/s. If the contact (during simulation) is assumed to be slip, the phase velocity decreases approximately by half, down to 900 m/s. This can be explained by the fact that, in the case of the slip contact, even the asymmetric guided wave mode *F*(1,1)-like propagates separately in each wire.

Therefore, in the case of the slip contact, there is no possibility to generate the *F*(1,1)-like mode along the overall structure of a multi-wire rope using the excitation (normal force) from the top effectively. In the case of the solid contact, the asymmetric mode *F*(1,1)-like corresponds to the fundamental mode *F*(1,1)-like of a much thicker simulated structure compared with the case of the slip contact.

Those results coincide well with the dispersion curves obtained using the analytical technique, which shows that the phase velocity of the $F(1,1)$ mode in a single wire should be around 900 m/s at the frequency of 60 kHz (Figure 2).

3. FINITE ELEMENT SIMULATION OF GUIDED WAVE PROPAGATION IN 3D WIRE ROPE MODEL

The penetration of the propagating guided waves into the structure of the wire rope mainly depends on boundary conditions between the particular wires. During the finite element simulations it is usually assumed that the contact between the wires is solid to simplify the complexity of the numerical model. In this case, the surface-tangential component of the particle velocity can propagate throughout the structure of the rope. However, it is common that the multi-wire ropes are oiled to increase the flexibility and reduce the wear and abrasion. In this case, the penetration of guided waves may be quite different.

To investigate the effects of guided wave propagation and penetration into deeper layers of the multi-wire rope, the finite element simulations of the straight 3D structure were carried out. For investigations the multi-wire rope with the polypropylene core had been selected. The length of the wire rope was equal to 0.7 m. The simulations were performed for two separate cases. In Case A, the acoustic contact between separate straight strands was assumed to be solid and had the properties of steel. In Case B, the thin layer of a new material named as "elastic fluid" with the shear wave velocity close to zero was introduced between the separate straight strands to simulate a slip contact. The 3D views of the wire rope models for the Case A and the Case B are presented in Figure 11.

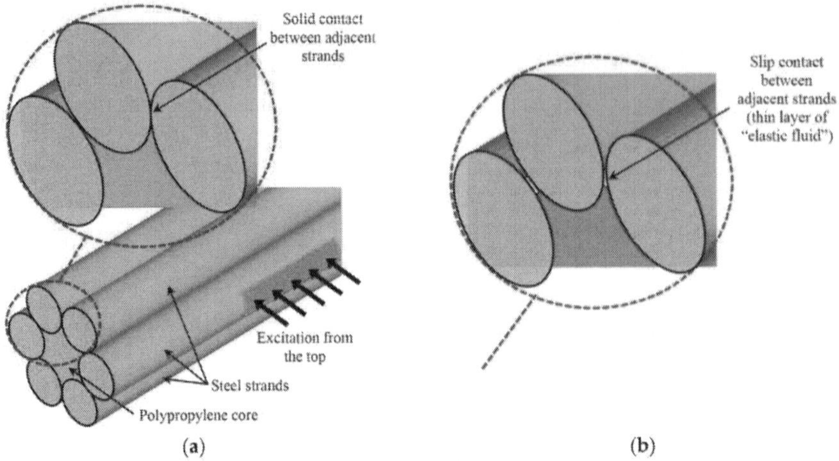

Figure 11. The 3D view of the multi-wire rope FE models in the case of a solid (**a**) and a slip (**b**) contact between neighboring strands.

The finite elements of the wire rope numerical model were created using a free mesh. The SOLID45 finite elements having eight degrees of freedom (translation in the *x*, *y*, and *z* directions) were used for meshing of the analyzed structure. The average spatial size of the elements was 2 mm for the steel region and 1.5 mm for the polypropylene. For the simulation of the layer of the "elastic fluid", the SOLID45 elements with the average spatial size of 1 mm were used. The number of nodes per wavelength of the slowest $F(1,1)$-like mode at 30 kHz frequency was 25 for steel, 33 for the polymer, and 50 for the elastic fluid.

In both cases the guided waves were generated by applying a normal force to the top surface of a separate strand. The excitation region had the dimensions of 15 mm × 10 mm and was placed 282 mm away from the nearest edge of the wire rope. This distance is equal to a single full whorl of the particular strand around the whole untwisted multi-wire rope.

In both cases the excitation pulse had the Gaussian envelope of seven periods and the frequency of 30 kHz. The bandwidth of the excitation pulse at the −6 dB level was 4.5 kHz. The Newmark time integration scheme was applied to obtain the numerical solution. The integration step in the time domain was 1.25 μs, which is 1/27 of the period at the 30 kHz central frequency. The spatial distributions of particle velocity components (axial v_z and perpendicular to the axis of the rope v_x) at different time instants were obtained using numerical integration. The C-scan images of the particle velocity distributions across the edge of the wire rope were created for the cases A and B. The C-scan images were created by selecting the peak-to-peak amplitude of the waveform at each available node of interest at time instants 55 μs, 81 μs, and 117 μs. Some further investigations demonstrated that the C-scan across the edge of the wire rope fully describes the spatial distribution of the wave amplitude.

The C-scan images of the axial v_z component (cross-section plane y-x) of the particle velocity obtained across the selected plane in the cases of a solid (case A) and a slip (case B) contacts are presented in Figure 12a–f. Please note that the amplitude scales in the C-scan images are different at different time instants and are increasing in time. The amplitudes at the time instant 117 μs are approximately 100 times larger than at 55 μs. The axial v_z component was selected for analysis in order to verify the obtained results by relevant experiments described further.

In the case of the solid acoustic contact mainly the flexural-like mode $F(1,1)$, which propagates forward, is strongly expressed within the overall cross-section of the multi-wire rope (Figure 12a,c,e). In the case of the slip acoustic contact the$F(1,1)$-like mode propagates along the single strand only (Figure 12b,d,f).

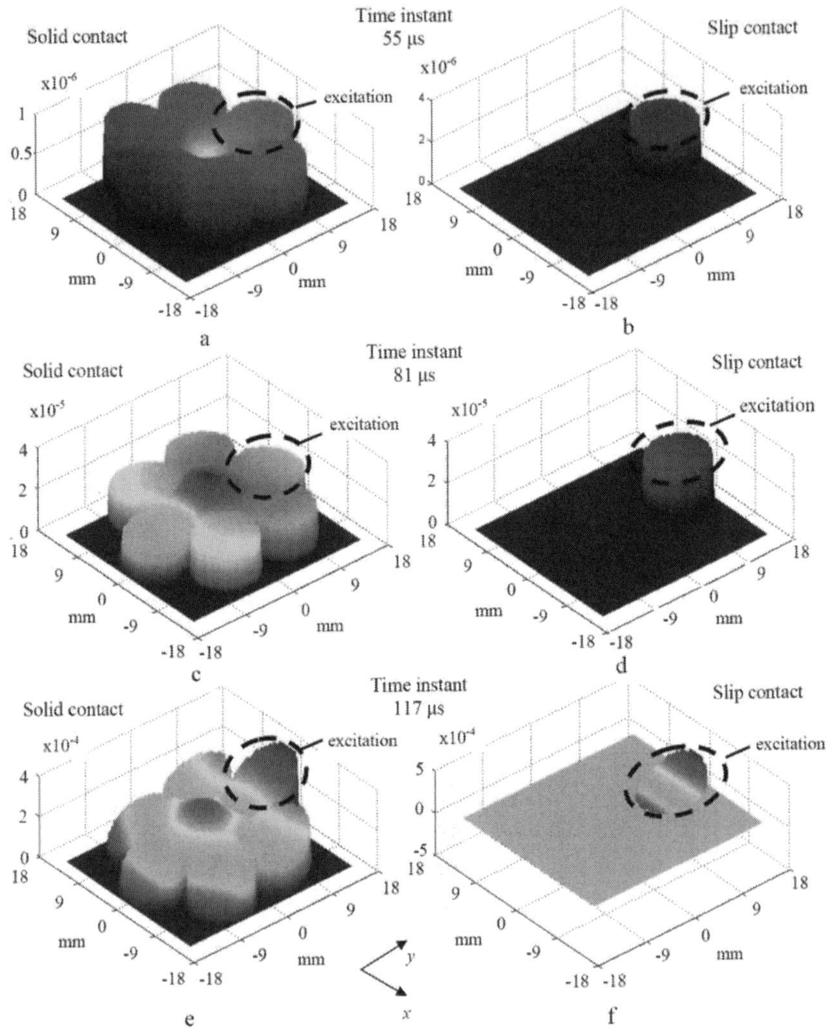

Figure 12. 3D representation of the axial component (v_z) of the particle velocity (cross-section plane *y*-*x*) at different time instants and acoustic contacts (solid or slip) between adjacent strands: time instant 55 µs ((**a**) solid contact, (**b**) slip contact), time instant 81 µs ((**c**) solid contact, (**d**) slip contact) and time instant 117 µs ((**e**) solid contact, (**f**) slip contact). The excitation was applied to the strand which is marked by a dashed line.

4. EXPERIMENTAL INVESTIGATION OF A MULTI-WIRE ROPE

From the numerical simulations it follows, that the penetration depth of the propagating UGW (mainly the flexural-like mode $F(1,1)$) into the internal structure of a multi-wire rope depends on the type of the acoustic contact between single wires and strands. The objective of the experimental investigation was to verify the effects of UGW propagation discovered by the FE simulation and to determine whether UGW propagate along the whole structure of wire rope or just in separate, externally-located strands.

The main purpose of the experiment was to reconstruct the wave penetration profile across the edge of the mock-up sample of the multi-wire rope. For the experimental investigation a composite rope with overall diameter of 35 mm and polymer core was selected (Figure 1).

In order to get the spatial distribution of the axial component amplitude across the edge of the multi-wire rope the air coupled technique was chosen. The immersion technique in this case is not suitable since the main goal of this experiment was to investigate the contact between separate strands. The water ingress can change the boundary conditions between the adjacent strands which would lead to inaccurate final results. The contact measurement technique was not suitable since it requires a stable acoustic force within the whole cross-section of the rope.

The experimental investigation of the wire rope was performed using the "Ultralab" ultrasonic measurement system, which had been developed at the Ultrasound Institute of Kaunas University of Technology. The experimental set-up for investigation of the multi-wire rope is presented in Figure 13. The ultrasonic guided waves were excited using the low frequency ($f = 30$ kHz, with a bandwidth of 20 kHz at the −6 dB level) ultrasonic transducers operating in the thickness mode (in order to create the normal force) and pitch-catch configuration. The transmitting transducer was attached to the surface of the particular strand using the specially developed spring-type adjusters. The contacting area of the conical delay line of the actuator of a concave shape was 134 mm^2 (projection of the contact area is the ellipse with the major radius of 8.6 mm and the minor radius of 5.8 mm). The transmitter was placed 272 mm away from the edge of the rope, which was equal to a single full whorl of the particular strand in the multi-wire rope. This distance is 10 mm shorter in comparison to the distance used in the finite element simulation (282 mm) as during the experiments the helical wire rope was used. To increase the transmission of the excitation force to the specimen a special conical delay lines with a concave contact surface were used. The transmitter was driven by the seven-cycle, 600 V tone burst with the central frequency of 30 kHz.

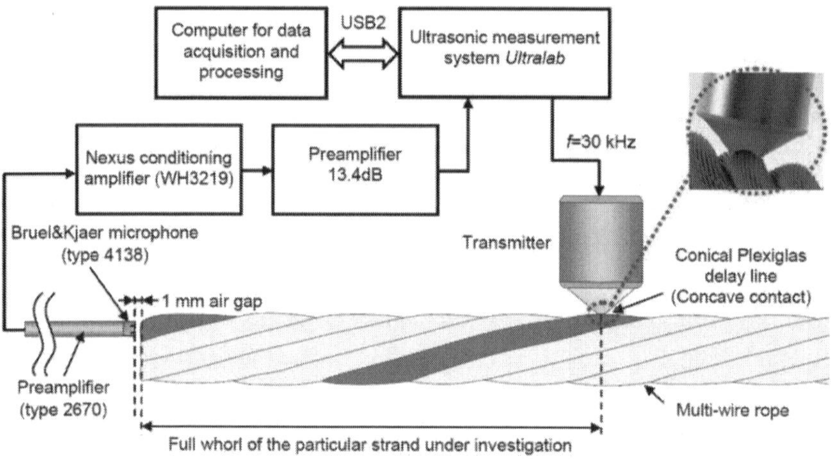

Figure 13. The experimental setup for the investigation of a guided wave propagation in a multi-wire rope.

The amplitude distribution across the edge of the rope was obtained by spatial scanning of a high frequency Brüel & Kjær air-coupled microphone (type 4138, Brüel & Kjær, Nærum, Denmark, frequency range: 6.5 Hz to 140 kHz, dynamic range: 52.2–168 dB, sensitivity 1 mV/Pa). The microphone was connected to the Falcon Range preamplifier (type 2670, Brüel & Kjær, Nærum, Denmark) and the Nexus conditioning amplifier (type WH3219, Brüel & Kjær, Nærum, Denmark). The diameter of the active region of the microphone was 3.175 mm, which is close to the diameter of a single wire of the rope. Before the experiments, the edge of the rope had been polished to get a precisely flat surface. The microphone was placed 1 mm from the edge of the rope, leaving as small air gap between the receiver and the specimen as possible. The microphone was operating in the near field, as the calculated near field length, taking into account the diameter of the particular strand and the wavelength of the longitudinal mode in the air, was equal to 2.22 mm. In general, the microphone makes it possible to register the sound pressure variations in air, across the edge of the multi-wire rope.

The sound pressure can be related to the particle velocity by the equation:

$$p(t) = \omega \times \xi \times Z = \frac{a(t) \times Z}{\omega} = \rho \times c \times v(t) \qquad (3)$$

where ω is the angular frequency (rad/s), ξ is the particle displacement (m), Z is the acoustic impedance of air (N·s/m^3),$a(t)$ is the particle acceleration (m/s^2), ρ is the density of air (kg/m^3), c is the speed of sound (m/s) in air, and $v(t)$ is the particle velocity (m/s).

During the experiment, the peak-to-peak amplitudes of the axial component of the registered waveforms of the $F(1,1)$-like modes were measured within time interval of 0–250 μs. The selected relatively short time interval enabled us to distinguish between the signals propagating in the structure of the wire-rope and those which propagate directly to the microphone through the air gap. The experimentally-obtained distribution of the peak amplitudes of axial components across the tip of the multi-wire rope is presented in Figure 14. It is possible to conclude that the acoustic contact between adjacent strands of the real multi-wire rope is close to the slip type, according to the observed effect that UGW propagate mainly along the same strand on which the normal type excitation was applied. It corresponds well to the results obtained by the numerical FE simulation (Figure 12).

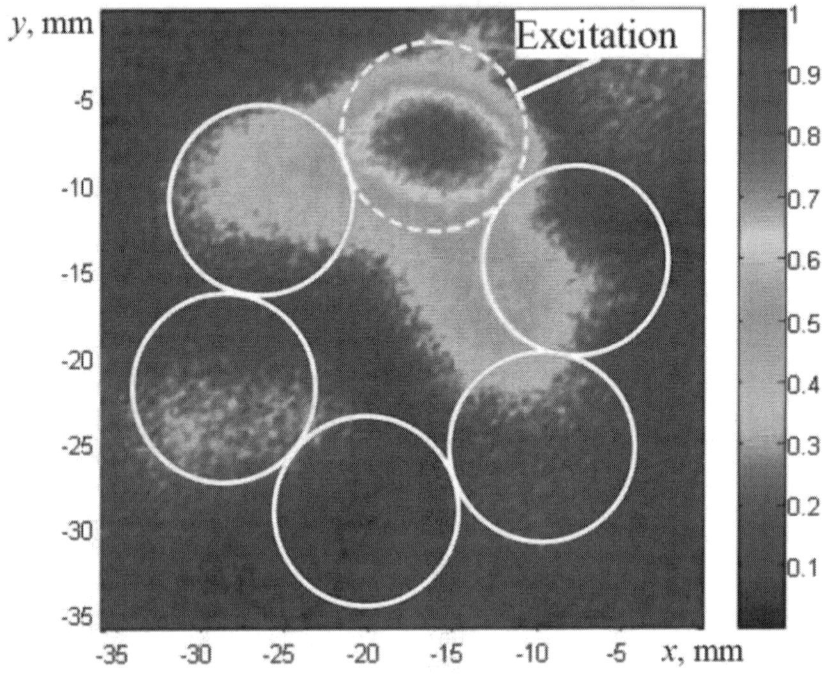

Figure 14. The experimentally-obtained C-scan image of the peak amplitudes of the axial components guided waves received across the cross-section of the multi-wire rope using the air coupled measurement technique. The excitation from the top was applied to the strand which is marked by a dashed line.

Previous investigations performed by contact type measurements of peak amplitudes of UGW (the $F(1,1)$-like mode) around the same strand and different strands proved that the highest amplitude of the received signal is registered when excitation and reception are performed on the same strand as well [23].

5. CONCLUSIONS

The important question regarding the development of the UGW-based testing technique has been the estimation of the real boundary conditions (solid or slip) between neighboring wires or strands and the real depth of penetration of UGW into the internal structure of the multi-wire rope. The propagating modes (mainly the $L(0,1)$-like and the $F(1,1)$-like) of UGW along the composite multi-wire rope have been identified by modelling using analytical, SAFE, and FE techniques. It has been estimated that some parts of the composite multi-wire rope (e.g., strands) may be replaced by a solid rod of the corresponding diameter. Otherwise, dimensions of the model are limited by a lack of computational resources and memory overflow.

During simulation of the row consisting of five wires placed on each other the obtained results have demonstrated that in the case of the edge type excitation (parallel to the axis of the wire) and the solid acoustic contact between adjacent wires, mainly the $L(0,1)$-like mode is generated. However, in the case of the solid contact, the $F(1,1)$-like mode is also generated due to mode conversions at each junction line between neighboring wires. In the case of excitation from the top by a normal force (perpendicular to the axis of the wire being excited), mainly the $F(1,1)$-like mode is generated for both types of the acoustic contact (slip and solid). The weak trace of the $L(0,1)$-like mode is also visible due to mode conversions.

The influence of different excitation regions and types of the acoustic contact between neighboring strands on the propagating UGW modes has been investigated using 3D FE modelling. Simulation results have demonstrated that in the case of the slip contact between adjacent strands and using the excitation from the top by a normal force the asymmetric-like mode $F(1,1)$ propagates separately in each strand. Therefore, in the case of the slip contact the possibility to generate efficiently the $F(1,1)$-like mode along the overall structure of a multi-wire rope is limited.

From the performed numerical modelling and relevant experiments it follows that the acoustic contact between adjacent strands of the real multi-wire rope is close to slip, according to the observed effect that UGW propagate mainly along the same strand on which excitation by a normal force has been applied. Investigations have demonstrated also that even in the case of the slip contact between wires the UGW can be used for integrity analysis (e.g., non-destructive testing) of particular strands of multi-wire ropes and to detect the defective ones.

The industrial demand for this work covers application of the developed ultrasonic pitch-catch testing technique, experimental set-up and special contact-type transducers to identify the particular defective strands in a multi-wire rope.

ACKNOWLEDGMENTS

The authors would like to thank PhD student A. Jankauskas for provided help with experiments.

AUTHOR CONTRIBUTIONS

Renaldas Raisutis, Rymantas Kazys and Liudas Mazeika conceived and designed the experiments, contributed in writing; Vykintas Samaitis and Egidijus Zukauskas performed numerical simulations, experiments and data analysis.

CONFLICTS OF INTEREST

The authors declare no conflict of interest.

REFERENCES

1. Yamasaki, T.; Tamai, S.; Hirao, M. Optimum excitation signal for long-range inspection of steel wires by longitudinal waves. *NDT E Int.* **2001**, *34*, 207–212.

2. Liu, Z.; Zhao, J.; Wu, B.; Zhang, Y.; He, C. Configuration optimization of magnetostrictive transducers for longitudinal guided waves inspection in seven-wire steel strands. *NDT E Int.* **2010**, *43*, 484–492.

3. Cheong, Y.M.; Kim, S.; Jung, H.K. Effect of an alternating excitation and the applicability of a magnetostrictive transducer for a long-range guided waves inspection. In Proceedings of the 12th A-PCNDT 2006—Asia-Pacific Conference on NDT, Auckland, New Zealand, 5–10 November 2006; pp. 1–6.

4. Laguerre, L.; Treyssede, F. Non-destructive evaluation of seven-wire strand using ultrasonic guided waves. *Eur. J. Environ. Civ. Eng.* **2011**, *15*, 487–500.

5. Lanza di Scalea, F.; Rizzo, P.; Seible, F. Stress measurement and defect detection in steel strands by guided stress waves. *J. Mater. Civ. Eng.* **2003**, *15*, 219–227.

6. Kwun, H.; Bartels, K.A.; Hanley, J.J. Effects of tensile loading on the properties of elastic-wave propagation in a strand. *J. Acoust. Soc. Am.* **1998**, *103*, 3370–3375.

7. Laguerre, L.; Aime, J.C.; Brissaud, M. Magnetostrictive pulse-echo device for non-destructive evaluation of cylindrical steel materials using longitudinal guided waves. *Ultrasonics* **2002**, *39*, 503–514.

8. Rizzo, P.; Di Scalea, F.L. Wave propagation in multi-wire strands by wavelet-based laser ultrasound. *Exp. Mech.* **2004**,*44*, 407–415.

9. Treyssede, F. Elastic waves in helical waveguides. *Wave Motion* **2008**, *45*, 457–470.

10. Treyssede, F.; Laguerre, L. Wave propagation in helical multi-wire cables. In Proceedings of the 7th European Workshop on Structural Health Monitoring, Nantes, France, 8–11 July 2014; pp. 2044–2051.

11. Raišutis, R.; Kažys, R.; Žukauskas, E.; Mažeika, L.; Vladišauskas, A. Application of ultrasonic guided waves for non-destructive testing of defective CFRP rods with multiple delaminations. *NDT E Int.* **2010**, *43*, 416–424.

12. Gaul, L.; Sprenger, H.; Schaal, C.; Bischoff, S. Structural health monitoring of cylindrical structures using guided ultrasonic waves. *Acta Mech.* **2012**, *223*, 1669–1680.

13. Baltazar, A.; Hernandez-Salazar, C.D.; Manzanares-Martinez, B. Study of wave propagation in a multiwire cable to determine structural damage. *NDT E Int.* **2010**, *43*, 726–732.

14. Treyssede, F.; Laguerre, L. Investigation of elastic modes propagating in multi-wire helical waveguides. *JSV* **2010**, *329*, 1702–1716.

15. Hayashi, T.; Rose, J.L. Guided wave simulation and visualization by a semi-analytical finite element method. *Mater. Eval.* **2003**, *61*, 75–79.

16. Hayashi, T. Guided wave animation using semi-analytical finite element method. In Proceedings of the 16th World Conference on NDT, Montreal, QC, Canada, 30 August–3 September 2004; pp. 1–8.

17. Hayashi, T.; Song, W.L.; Rose, J.L. Guided wave dispersion curves for a bar with an arbitrary cross-section, rod and rail example. *Ultrasonics* **2003**, *41*, 175–183.

18. Bartoli, I.; Phillips, R.; Lanza di Scalea, F.; Salamone, S.; Coccia, S.; Sikorsky, C.S. Load monitoring in multiwire strands by interwire ultrasonic measurements. *Proc. SPIE* **2008**, *6932*, 1–12.

19. Marzani, A.; Viola, E.; Bartoli, I.; Lanza di Scalea, F.; Rizzo, P. A semi-analytical finite element formulation for modeling stress wave propagation in axisymmetric damped waveguides. *JSV* **2008**, *318*, 488–505.

20. Laguerre, L.; Grimault, A.; Deschamps, M. Ultrasonic transient bounded-beam propagation in solid cylinder waveguide embedded in solid medium. *J. Acoust. Soc. Am.* **2007**, *121*, 1924–1934.

21. Beard, M.; Lowe, M.J.S.; Cawley, P. Ultrasonic guided waves for inspection of grouted tendons and bolts. *J. Mater. Civ. Eng.* **2003**, *15*, 212–218.

22. Kazys, R.; Butenas, G. Investigation of time-reversed ultrasonic waves focusing in multilayered media. *Ultragarsas***2008**, *63*, 43–48.

23. Raisutis, R.; Kazys, R.; Mazeika, L.; Zukauskas, E.; Samaitis, V.; Jankauskas, A. Ultrasonic guided wave-based testing technique for inspection of multi-wire rope structures. *NDT E Int.* 2014, *62*, 40–49.

CHAPTER 5

Ultrasonic propagation in finite-length granular chains

D.A. Hutchins , J. Yang[a], O. Akanji, P.J. Thomas, L.A.J. Davis, S. Freear, S. Harput, N. Saffari[c], P. Gelat

ABSTRACT

A narrowband ultrasound source has been used to generate solitary wave impulses in finite-length chains of spheres. Once the input signal is of sufficient amplitude, both harmonics and sub-harmonics of the input frequency can be generated as non-linear normal modes of the system, allowing a train of impulses to be established from a sinusoidal input. The characteristics of the response have been studied as a function of the physical properties of the chain, the input waveform and the level of static pre-compression. The results agree with the predictions of a theoretical model, based on a set of discrete dynamic equations for the spheres for finite-length chains. Impulses are only created for very small pre-compression forces of the order of 0.01 N, where strongly non-linear behaviour is expected.

1. INTRODUCTION

The motivation for this study was the need for high amplitude, wide bandwidth ultrasonic impulses for biomedical applications. High amplitudes can be derived from ultrasonic horns, but these are of narrow bandwidth. Granular chains could, in principle, increase the available bandwidth, due to the strongly non-linear Hertzian interaction between neighbouring particles [1]. The dynamic response of these systems to impulsive input signals is governed by nonlinear, dispersive and dissipative effects, leading to the formation of solitary waves that travel along the chains [2], [3] and [4]. Nesterenko has shown that characteristic solitary waves are predicted in an analytical solution using the long wave approximation [1] and [5]. Strongly nonlinear waves have also been observed in a chain of Teflon beads [6] and in double power-law materials [7], as well as in granular dimer chains [8].

The tunability of solitary wave properties has been demonstrated [9]. It is now known that the speed of solitary waves along a chain can be changed using an applied pre-compression force, provided the conditions of the chain are appropriate. This latter effect was used to demonstrate that a nonlinear acoustic lens with a tuneable focus can be produced [10], leading to the interesting concept of sound bullets. These sound bullets are established from an appropriate superposition of the individual output signals from a set of neighbouring Hertzian chains. These previous studies [9], [10] and [11]demonstrated that piezoelectric actuation could be used to generate solitary waves in columnar chains of steel ball-bearings under small pre-compression (F_0). The resultant effects are dependent on the characteristics of the applied transient force (F_m), the diameter of the spheres, and the relative values of F_m and F_0. If $F_m \approx F_0$, weakly non-linear behaviour results, whereas if $F_m \gg F_0$, propagation along the chain is highly non-linear.

Most of the experimental work to date has used relatively low ultrasonic frequencies and large spheres with diameters of the order of up to 10 mm. The cut-off frequency for solitary waves limits the upper frequency that can propagate in chains [1]; hence, higher frequencies will need smaller spheres. In addition, little work seems to have been performed experimentally into establishing whether higher frequencies and non-impulsive waveforms could be used in chains of finite length. Of particular interest in this context is the establishment of resonances within such chains. Various papers have described the concept of Non-linear Normal Modes (NNMs) of vibration, which are expected to exist in such systems [12], [13] and [14]. Jayaprakash et al. [12] studied such chains with fixed boundary conditions, and predicted a set of NNMs which resemble a traveling wave propagating backwards and forwards through the finite granular chain. These exist within allowed frequency bands (known as propagation bands), which have been observed in a two-sphere experimental system [14]. NNMs are expected to be very sensitive to the input conditions, in terms of the nature of F_m and F_0, the types of spheres used and the boundary conditions. Recent work by the current authors has investigated finite-length chains using ultrasound of a higher frequency and smaller spheres [15]. It was observed that for a given input frequency, impulses evolved once the input amplitude exceeded a specified threshold value for a certain length chain. The present paper explores this process in much greater detail, and investigates the factors which govern the creation of these impulses. Discrete dynamic equations of motion for chains of spherical oscillators are also used to predict the signal development and propagation in such chains, for comparison to the corresponding experimental results.

2. APPARATUS AND EXPERIMENT

The experimental arrangement used is shown in Fig. 1. Chains consisting of 1 mm diameter chrome-steel spheres, with lengths varying between 2 and 14 spheres, were placed within cylindrical holders fabricated from acrylic resin. The holders were manufactured individually from acrylic resin to suit each length of chain, using micro-stereolithography (a form of additive manufacturing). The spheres were placed horizontally within the holder so as to just touch each other, minimising the initial static pre-compression force (F_0). The first sphere was excited harmonically by a longitudinal ultrasonic horn, operating at 73 kHz, which was driven by a high voltage tone-burst signal using an Agilent 33120A function/Arbitrary waveform generator and a power amplifier. Both the horn and the chains of spheres were clamped rigidly onto an optical translation stage, and a micrometer used to position the horn against the first sphere with as little force as possible. The last sphere in the chain was held in place using an annular aperture, allowing detection of the particle velocity waveform at the output via a Polytec laser vibrometry system. The tone-burst duration and amplitude of the drive voltage signal could both be adjusted. The ultrasonic motion of the horn tip at full power, for drive voltage tone-bursts of 20 cycles and 45 cycles in duration, is shown in terms of its waveform and frequency spectrum in Fig. 2(a) and (b) respectively. Both show a prominent peak at 73 kHz, with only a small amount of harmonic content from the vibration of the horn.

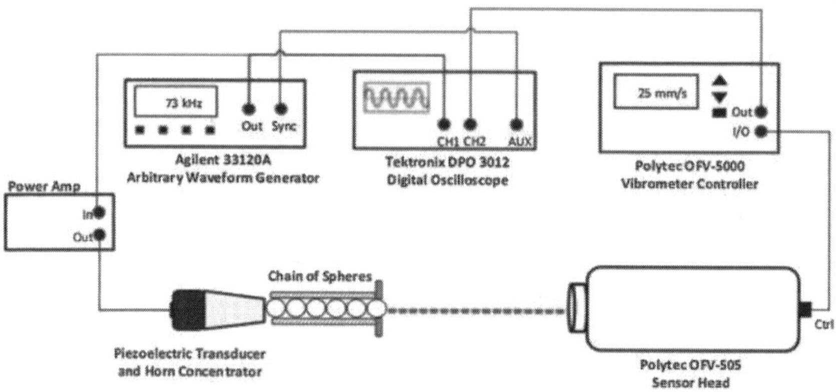

Figure. 1. Experimental arrangement used to detect signals generated by the ultrasonic horn in the chain of spheres.

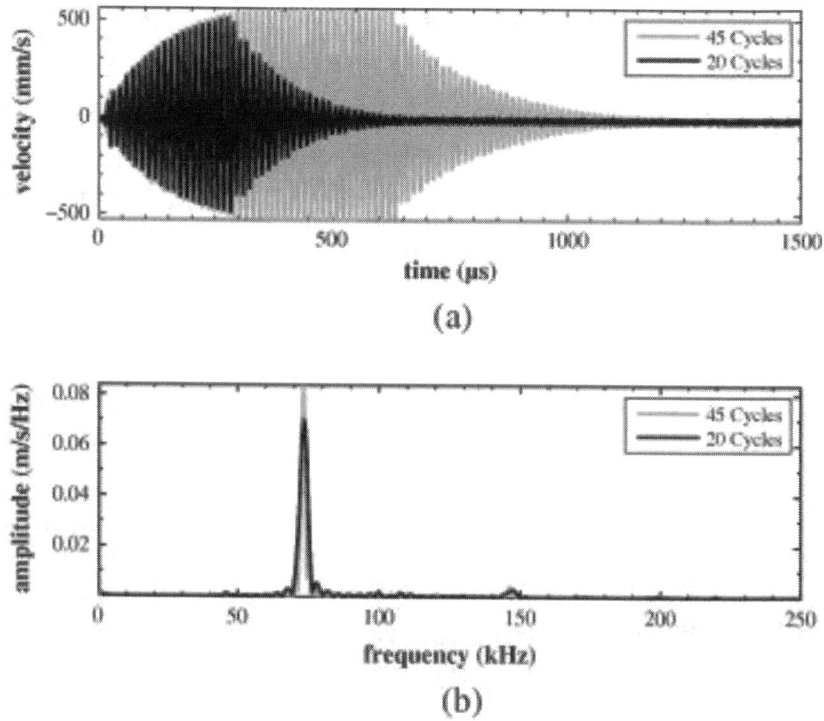

Figure. 2. (a) Waveforms and (b) spectra at the horn tip at maximum power when driven by voltage tone-bursts with durations of 20 and 45 cycles.

3. DISCRETE THEORETICAL MODELLING OF WAVE PROPAGATION IN CHAINS

Here we model the acoustic wave transmission in a granular medium of finite length, contained between the ultrasonic horn and a reflecting wall, so as to understand the relationship between the input and output signals and the chain structure. This analysis also has to consider the boundary conditions at either end of the chain, as these have an effect on the resulting behaviour in a resonant system. The theoretical analysis also allows prediction of the effects resulting from changes in the static pre-compression force F_0.

The discrete dynamic equations of motion [15] that describe the types of granular chains studied here assume that the spheres have an identical mass (m) and radius (R). The first sphere is harmonically excited by an ultrasonic transducer with a displacement that can be calculated from the particle velocity waveform measured experimentally at the horn tip (see Fig. 2). The first sphere in the chain is in contact with the planar face of the horn; the last sphere is assumed to be in contact with a fixed planar wall (although in practice this has an opening to allow measurement of the vibration of the last sphere in the chain). The planar surfaces of these two boundaries are, in fact, modelled as spheres with very large radii.

The model predicts the displacements of the centres of consecutive spheres (u_1, u_2, u_3, u_N), for a particular input displacement waveform (u_0) from the horn tip. The experimental waveforms are observed to exhibit energy loss with elapsed time; hence, a linear viscous damping coefficient (λ) is used within the dynamic equations of motion, so that energy dissipation is included. This also helps the model to become more stable, a phenomenon observed by other authors [16]. Damping is only relevant when the spheres are in contact; if this is not the case, then a damping term is not needed (as previously described by Lydon et al. [14]). This is dealt with via the Heaviside function Hincluded within the equations. The effect of a static pre-compression force F_0 is modelled by computing an initial static displacement of the centre of each sphere in the chain from its zero-force state. This differs at the two ends of the chain from that present for intermediate spheres, because of boundary conditions. The resulting starting displacements are denoted as δ_{0l} (between the horn and the first sphere), δ_0 (between intermediate spheres), and δ_{0r} (between the last sphere and the end wall) respectively.

As will be demonstrated below, the exact conditions of the chain have to modelled carefully, as the nonlinear normal modes that are created are sensitive to this. It is thus necessary to define effective Young's moduli associated with contact interactions between the horn and the first sphere, between intermediate spheres, and between the last sphere and the end wall (θ_l, θ_m and θ_r respectively). These quantities have values which depend on the material properties of the spheres, according to

$$\frac{1}{\theta_l}=\frac{1-\nu_l^2}{E_l}+\frac{1-\nu_s^2}{E_s} \quad \theta_m=\frac{E_s}{1-\nu_s^2} \quad \frac{1}{\theta_r}=\frac{1-\nu_r^2}{E_r}+\frac{1-\nu_s^2}{E_s}, \tag{1}$$

where E_l and ν_l are the Young's modulus and Poisson ratio of the horn, E_r and ν_r are that of the far end wall, and E_s and ν_s those of the spheres themselves. The values ofδ_{0l}, δ_0 and δ_{0r} are then given by

$$\delta_{0l}=\left(\frac{3F_0}{4\sqrt{R}\theta_l}\right)^{2/3} \quad \delta_0=\left(\frac{3F_0}{\sqrt{2R}\theta_m}\right)^{2/3} \quad \delta_{0r}=\left(\frac{3F_0}{4\sqrt{R}\theta_r}\right)^{2/3}. \tag{2}$$

For the input sphere next to the horn, the equation of motion in terms of particle displacement in a chain containing N spheres is [15]:

$$\begin{aligned} m\frac{d^2u_1}{dt^2}=&\frac{2\sqrt{R}}{3}\left[2\theta_l(\delta_{0l}+u_0-u_1)^{3/2}-\frac{\theta_m}{\sqrt{2}}(\delta_0+u_1-u_2)^{3/2}\right]\\ &+\lambda\left(\frac{du_0}{dt}-\frac{du_1}{dt}\right)H(\delta_{0l}+u_0-u_1)-\lambda\left(\frac{du_1}{dt}-\frac{du_2}{dt}\right)H(\delta_0\\ &+u_1-u_2). \end{aligned} \tag{3}$$

For intermediate spheres in the chain, the relevant equation of motion is:

$$\begin{aligned} m\frac{d^2u_i}{dt^2}=&\frac{\sqrt{2R}}{3}\theta_m[(\delta_0+u_{i-1}-u_i)^{3/2}-(\delta_0+u_i-u_{i+1})^{3/2}]\\ &+\lambda\left(\frac{du_{i-1}}{dt}-\frac{du_i}{dt}\right)H(\delta_0+u_{i-1}-u_i)-\lambda\left(\frac{du_i}{dt}-\frac{du_{i+1}}{dt}\right)H(\delta_0\\ &+u_i-u_{i+1}). \end{aligned} \tag{4}$$

At the far end of the chain, the equation becomes

$$\begin{aligned} m\frac{d^2u_N}{dt^2}=&\frac{2\sqrt{R}}{3}\left[\frac{\theta_m}{\sqrt{2}}(\delta_0+u_{N-1}-u_N)^{3/2}-2\theta_r(\delta_{0r}+u_N)^{3/2}\right]\\ &+\lambda\left(\frac{du_{N-1}}{dt}-\frac{du_N}{dt}\right)H(\delta_0+u_{N-1}-u_N)-\lambda\frac{du_N}{dt}H(\delta_{0r}+u_N). \end{aligned} \tag{5}$$

Note that negative values of the force terms involving successive spheres in Eqs. (3),(4) and (5) imply that the relevant two neighbouring spheres have lost contact and the terms are then set equal to zero; additionally the damping term involving the Heaviside function H then becomes inactive, as explained above.

For comparison to the present experiments, numerical solutions of these equations were obtained using a technique developed for molecular dynamics, namely the Velocity Verlet method. This method has been used, for example, in the simulation of the dynamics of particles in suspensions [16]. It is particularly suited to discrete systems, such as the present situation involving the study of the motion of individual spheres. The relevant material parameters required for the calculations are given in Table 1. The value of λ required to get good agreement with experiment increased with the length of each chain, due to increased dissipation, over the range λ = 0.23–0.32 N s m^{-1} from 3 to 10 spheres in length. This damping is thought to arise primarily from dissipation caused by contact of spheres with the viscoeleastic polymer holder used to contain the spheres.

Table 1. Material constants used in the theoretical model.

Material	Young's modulus (GPa)	Poisson's ratio
Ultrasonic horn (steel)	$E_l = 201$	$\nu_l = 0.3$
Spheres (steel)	$E_s = 201$	$\nu_s = 0.3$
End wall (acrylic polymer)	$E_r = 2.45$	$\nu_r = 0.35$

4. EXPERIMENTAL RESULTS AND COMPARISON TO THEORY

Here, we report details of the experimental results, including the dependence of propagation along the chain on the input waveform characteristics. This has been performed for a series of chain configurations, in terms of chain length, sphere size, sphere material and end-wall conditions. Additional comparisons of experimental results to theoretical predictions allow a more in-depth analysis of the behaviour of the chains to be performed.

4.1. A chain of 6 steel spheres of 1 mm diameter

Experiments were performed initially on a chain of six chrome-steel spheres, and particle velocity waveforms recorded at the last sphere in the chain as the input was varied in terms of tone-burst duration and amplitude. Note that the vibrometer was also used to record the input signal from the horn tip in all experiments, so that the correct input waveform could be used for theoretical predictions to compare with a given experiment. These experiments were all performed using the minimum static pre-compression force (F_0) achievable in the apparatus of Fig. 1, with the horn just touching the first sphere. Comparisons between computations and experiments, to be discussed further below, revealed that the magnitude of this force is of the order of 0.01 N.

It was found experimentally that the output signal at the far end of the chain was highly dependent upon the characteristics of the input waveform, in terms of both amplitude and duration of the input signal at 73 kHz. Consider first the use of a 20-cycle voltage excitation signal into the horn (see Fig. 2 for the input velocity waveform from the horn at maximum power). Fig. 3 shows the results obtained for a 6-sphere chain as the input amplitude was gradually increased up to this maximum. Here we use v_m to represent the maximum pk–pk amplitude of the input velocity waveform for each measurement. At low amplitudes of the input (v_m = 87 mm s^{-1}), the output signal at the far end of the chain was small, with characteristics that were mainly those of the drive signal, although a small increase in the harmonic at 146 kHz indicated some limited non-linearity. As the input signal was then increased in amplitude, the time waveform started to change so as to incorporate a second feature with a period different to that of the input signal. This was associated with the appearance of additional regularly-spaced peaks in the frequency spectrum, at both higher and lower frequencies than the input at 73 kHz. As the magnitude of the input signal increased further, the new periodicity in the time waveform became more obvious, so that at the highest input amplitude, a set of distinct regularly-spaced impulses can be seen in the time waveform. This was associated with an increase in the prominence of the set of regularly-spaced peaks in the frequency spectrum. It is interesting to note that these frequency peaks were spaced at a value of 1/3 of that of the excitation frequency. They thus represent a set of two sub-harmonics, plus a set of harmonics, generated as a result of propagation along the chain. All the main frequency peaks are integer multiples of the lowest peak in the spectrum at 24.3 kHz. The periodicity of impulses in the time waveform was 41 μs, the reciprocal of this value. Note that at the higher input amplitudes (v_m = 937 mm s^{-1}), the two sub-harmonic peaks were of higher amplitude than any of the harmonics. This generation of distinct sub-harmonics is observed in our experiments to be a characteristic feature of such systems in situations where impulses appear in the time-domain waveform.

The experimental results shown in Fig. 3 can be compared to the predictions of theory, so as to gain greater insight into the processes at work. As stated above, the experimental arrangement is assumed to impose a small pre-compression force (F_0) onto the chain. At low input amplitudes (v_m = 87 mm s^{-1}), a weakly non-linear behaviour is exhibited in the experiment, with the enhancement of the first harmonic, indicating the presence of such a small pre-compression force. Such behaviour was also observed in the work of Lydon et al. [14] in their study of a two-sphere system. They indicated that experimental and numerical results did not agree quantitatively in the nominally zero pre-compression state; they then went on to analyse the situation more closely, and identified a small unintended pre-compression as the cause. In the present case, the addition of a small pre-compression force F_0 = 0.012 N was seen to produce good agreement between theory and experiment. The predicted waveform of Fig. 4 is now similar to that measured experimentally, in that the predicted signal is predominantly that of the input, modified by the addition of some harmonic content in the frequency spectrum. Hence a pre-compression force of this magnitude was thus assumed to exist in the modelling predictions to follow.

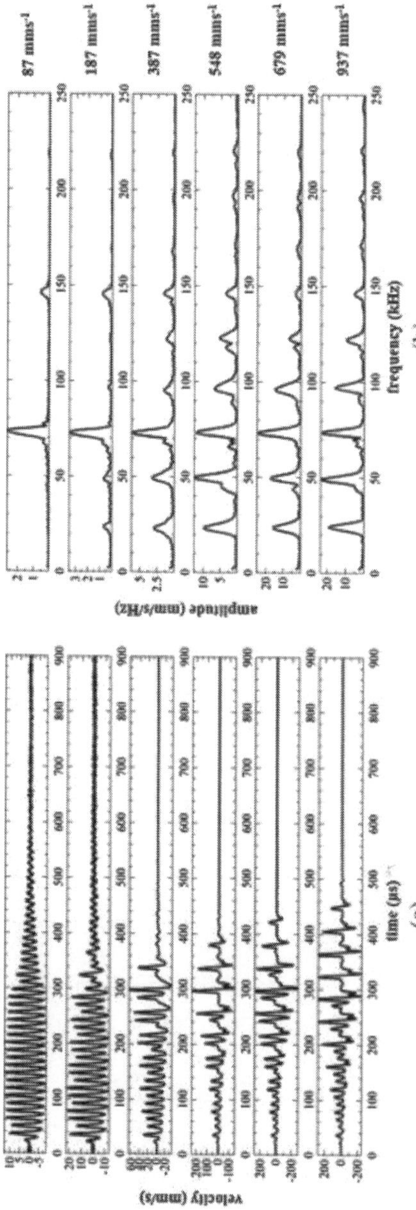

Figure. 3. (a) Waveforms and (b) frequency spectra of signals recorded at the output of a 6-sphere chain as the input amplitude was increased. The input was in the form of a tone-burst from the ultrasonic horn at 73 kHz, such as that shown in Fig. 2(a). The maximum pk–pk input amplitude (v_m) is given on the right in each case.

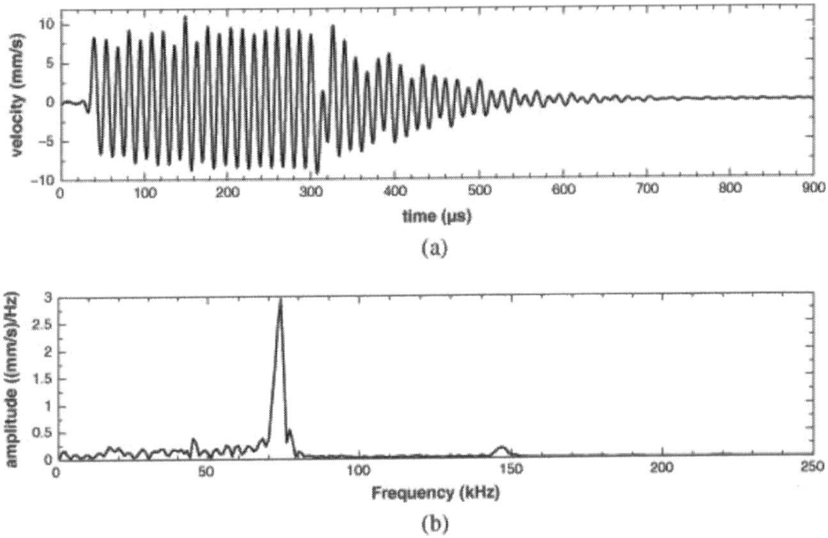

Figure. 4. Modelling results for the output signal from a 6-sphere chain, assuming a small input amplitude ofv_m = 87 mm s^{-1} from the ultrasonic horn, for a pre-compression force of $F_0 = 0.012$ N. (a) The predicted time waveform and (b) the corresponding frequency spectra.

The predicted waveform and corresponding frequency spectrum for an intermediate input amplitude of v_m = 378 mm s^{-1} (Fig. 5 (top)) contain features that agree with those seen experimentally, namely a form of transitional behaviour between the weakly nonlinear signals and the strong solitary wave pulses. Fig. 5 also shows predictions for the highest input amplitude (v_m = 937 mm s^{-1}), where the predicted motion at the end of the chain is a set of impulses with strong sub-harmonics in the spectrum, and has the main features seen in the experimental waveform of Fig. 3 at the highest input amplitude. The correct frequency components are predicted, as are the periodicity and general nature of the impulses in the time domain. There are minor differences in the magnitude of the frequency peaks between theory and experiment, but overall the agreement is very good.

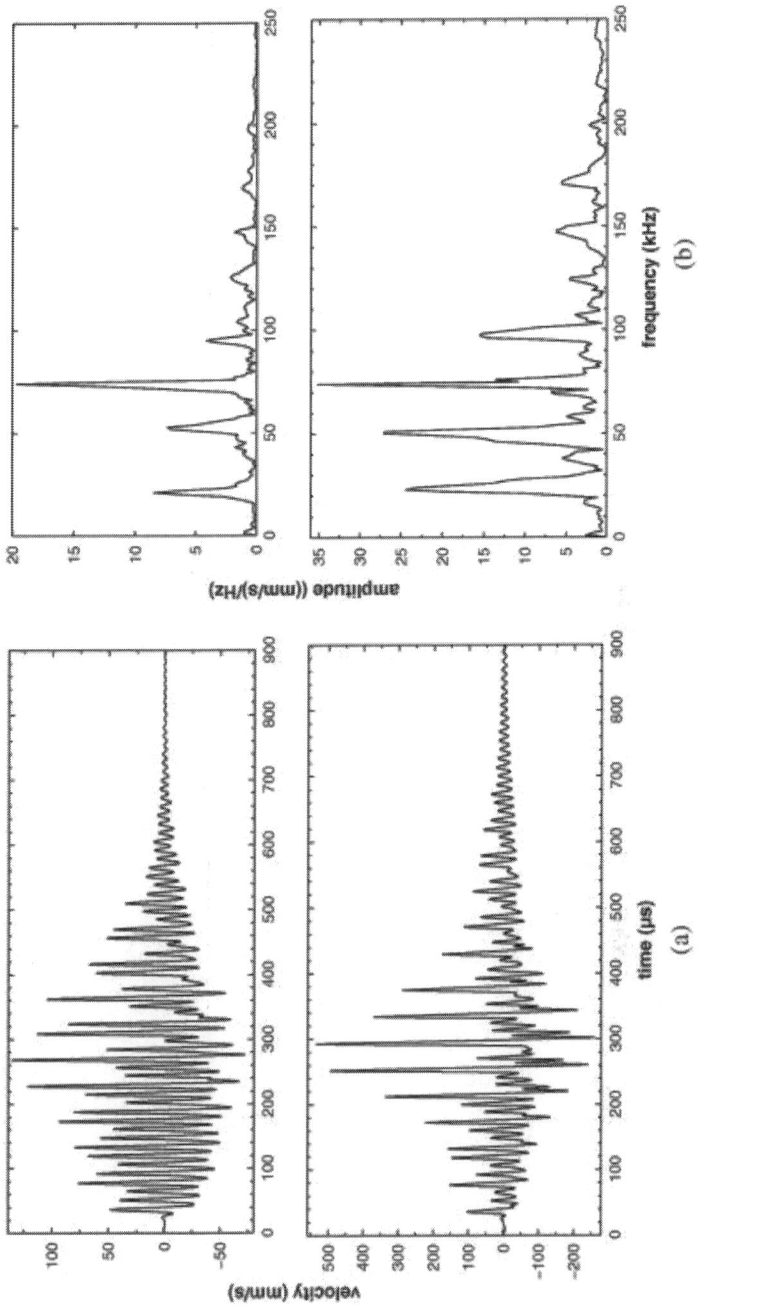

Figure. 5. Modelling predictions for a 6-sphere chain for both an intermediate amplitude of v_m = 378 mm s^{-1} (top) and at the maximum input amplitude of v_m = 937 mm s^{-1} (bottom). (a) The predicted time waveform and (b) the corresponding frequency spectrum in each case.

Experimentally, there was also a dependence on the number of cycles in the excitation waveform, as is demonstrated in Fig. 6 for 10, 20 and 30-cycle excitations. At 10 cycles, the impulses were not fully formed, and the spectral peaks were indistinct. The optimum result in terms of the creation of impulses occurred at 20 cycles. At longer inputs of 30 cycles, the amplitudes of both harmonics and sub-harmonics decreased relative to the main input frequency of 73 kHz, and the impulses became less distinct. This process continued for longer input signals. Note also that the measured pk–pk amplitude increased with longer excitation waveforms, a consequence of the additional energy being input into the spherical chain.

4.2. Results for different chain lengths of 1 mm diameter steel spheres

The above experiments were repeated for different chain lengths, ranging from 2 to 10 spheres. As in the case for 6 spheres, the output was a function of both the amplitude and the number of cycles in the waveform. The results for a 10-sphere chain of 1 mm diameter steel spheres are shown in Fig. 7, again as a function of input amplitude. It was found that the optimum result in terms of impulse formation for a 10-sphere chain occurred when the tone-burst excitation waveform was increased to be 45 cycles in length (an example was shown earlier in Fig. 2). In this case, the impulses were fully formed at the maximum input amplitude, and were seen to exist over a longer time duration than in the 6-sphere case, before again decaying away once the excitation was removed. Note that the input frequency of 73 kHz was a prominent peak for both chain lengths of 6 and 10 spheres, as would be expected. However, now the frequency spectrum of the longer chain contained three sub-harmonics, instead of the two seen for the 6-sphere case, and the lowest frequency peak was at a lower frequency of 18.25 kHz (*i.e.* 1/4 that of the input). The corresponding time between impulses was 54.8 μs. Note that this change in time separation between impulses is not in direct proportion to chain length, when a comparison is made to the 6-sphere case. This is not unexpected for such a highly-nonlinear system. It is also evident that the longer chain length has resulted in a wider bandwidth at the maximum input signal level, with sharper peaks in the time waveform, and an extended spectrum.

Modelling confirms the expected behaviour for this longer chain. The predictions are shown in Fig. 8 for the 45-cycle input at maximum amplitude. Again, the main features of the experimental waveform and spectrum are predicted – distinct impulses are predicted at the correct spacing in time, with the spectrum containing peaks separated by a value 1/4 that of the input frequency. Prominent sub-harmonics are present, as expected.

Comparison of the results for the two lengths of chain indicates some interesting features. The time-domain train of impulses has a regular periodicity in each case, but with a longer period for the longer chain. This is associated with a smaller separation of the equivalent regular frequency peaks in the frequency spectrum. In addition, all of the main peaks are multiples of the lowest frequency peak (1/3 and 1/4 of 73 kHz for the 6 and 10 sphere chains respectively). These represent some of the allowed NNMs of the chain. The first peak in each case represents the lowest-frequency NNM resonance permitted as a whole fraction (*i.e.* a sub-harmonic) of the input frequency for that chain length. This reciprocal of this frequency corresponds to the time interval between impulses in each time waveform. Both harmonics and sub-harmonics are generated, as they are all allowed resonances of the chain. Note that the input drive waveform will contain some energy at all frequencies in this spectrum, due to its finite time duration, even though this may be very small; hence, sub-harmonics can grow from this small seed energy level.

Experimentally, it is observed that the increase in the maximum pk–pk amplitude of the particle velocity measured at the output is not a linear function of the input amplitude. This is illustrated in Fig. 9 for chains containing 6 and 10 spheres. Initially, the output is approximately proportional to the input levels. However, at a certain point, the amplitudes grow more quickly, before reaching a saturation level, after which the amplitude appears to vary erratically. This behaviour is not dissimilar to that expected from finite-amplitude resonant systems in which sub-harmonic generation is involved, such as those seen in bubble dynamics [17] and [18].

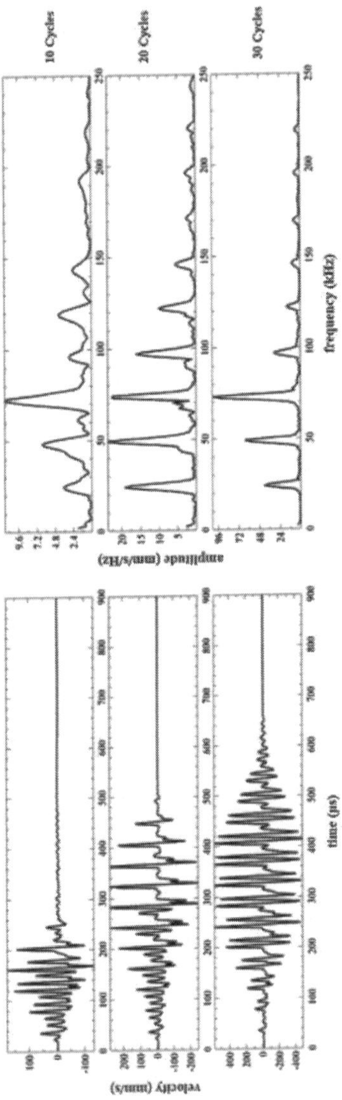

Figure. 6. Experimental waveforms showing the effect of changing the number of cycles (given on the right) in the drive waveform for a 6-sphere chain, at the maximum input amplitude of $v_m = 937$ mm s^{-1}. (a) Time waveform and (b) corresponding frequency spectrum.

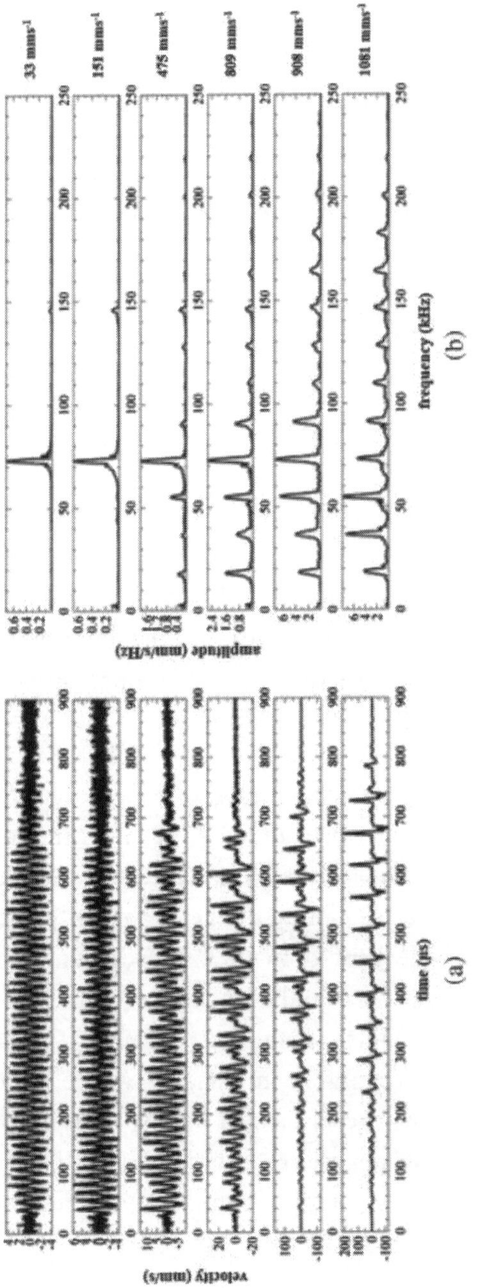

Figure. 7. (a) Waveforms and (b) frequency spectra of output signals recorded for a 10-sphere chain, using a 45-cycle tone-burst from the ultrasonic horn at 73 kHz (Fig. 2(a)). The maximum pk–pk input amplitude v_m is given on the right in each case.

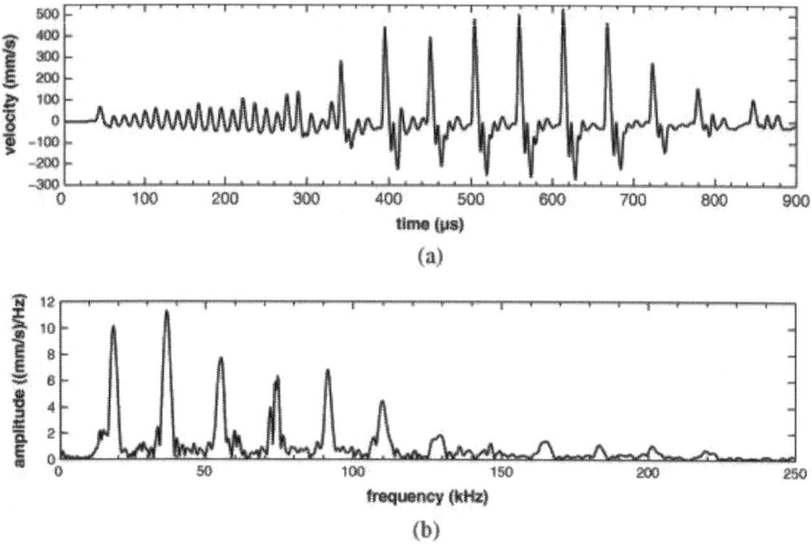

Figure. 8. Modelling results for a 10-sphere chain, assuming the 45-cycle time waveform of Fig. 2(b) as the input at the highest input amplitude from the horn at v_m = 1081 ms^{-1}(Fig. 2(b)). (a) Time waveform and (b) frequency spectrum.

The result obtained experimentally for a 3-sphere chain of 1 mm diameter spheres is shown in Fig. 10 (top). As might be expected, it falls within the pattern observed thus far – now there is only one sub-harmonic, at 1/2 the frequency of the input, with the corresponding repetition rate of the time-domain impulses being the reciprocal of this value. There is also only one distinct harmonic. The trends observed for the chains containing 6 and 10 spheres have thus continued – for the creation of impulses, a whole fraction of the input frequency needs to be present as the lowest peak in the spectrum, and all other significant peaks are multiples of this. In addition, the apparent propagation velocity of impulses within the chain decreases as the chain gets shorter.

Experiments have been performed for all chain lengths in the 2–14 range. The effects of resonance and the generation of a train of impulses for sphere diameters of 1 mm at an excitation frequency of 73 kHz were only observed in the case of 3, 6 and 10 spheres. This was irrespective of the amplitude or number of cycles in the input signal. This indicates that conditions were not well-matched for the creation of NNM resonances and pass-band behaviour in these chain lengths, for the size of steel spheres and input frequency used.

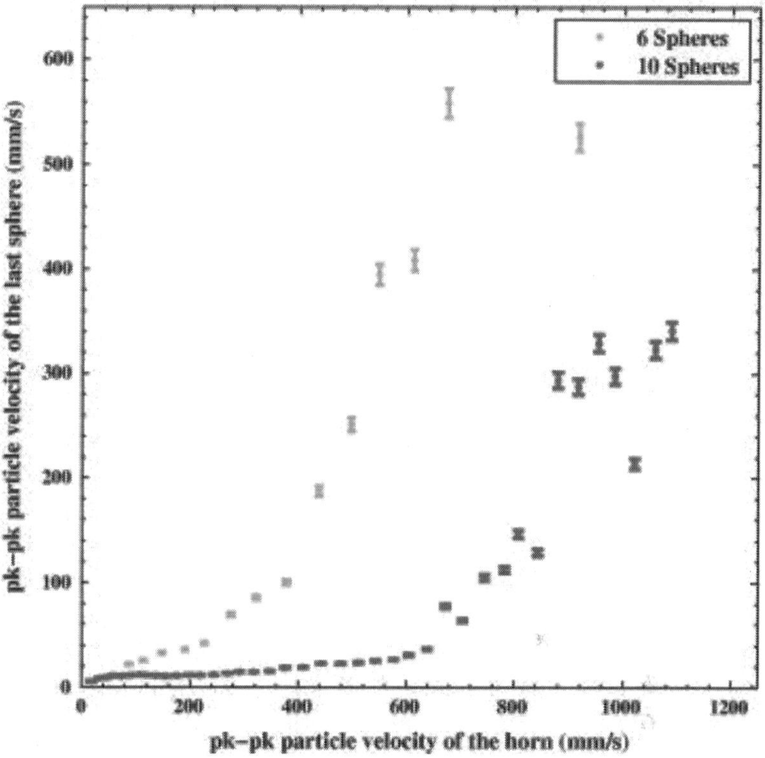

Figure. 9. Dependence of output pk–pk maximum amplitude on that of the input, as measured by the vibrometer in chains containing both 6 and 10 spheres.

4.3. Use of a different sphere material

Experiments were also conducted on 1 mm diameter thermoplastic polymer spheres, in this case polyoxymethylene (Delrin™), and tungsten carbide ceramic spheres of the same size. Table 2 compares their material properties to those of acrylic polymer (used to make the holder for the spheres) and the steel ball-bearings.

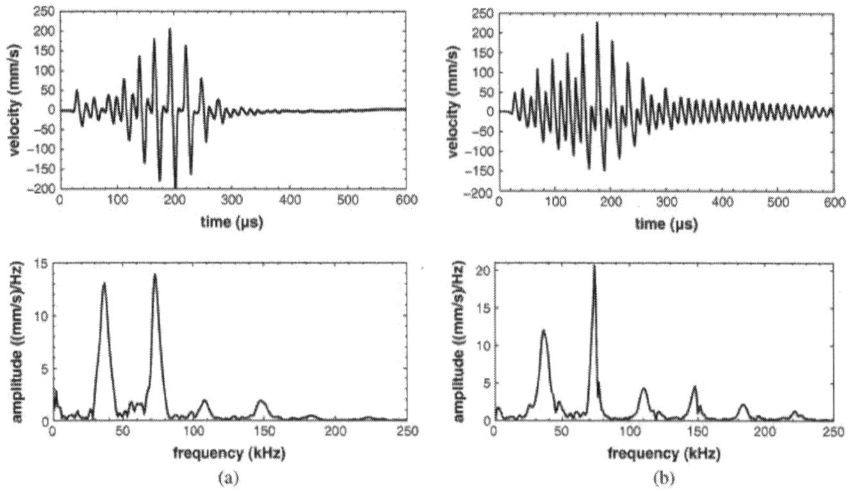

Figure. 10. Results for a 3-sphere chain of steel spheres of 1 mm diameter using an input waveform in each case of 10 cycles at 73 kHz and with $v_m = 634$ mm s^{-1}. (a) Experimental result and (b) theoretical prediction, showing waveform (top) and frequency spectrum (bottom) in each case.

It can be seen that Delrin differs from steel in being reasonably well-matched in terms of the acoustic impedance vale (Z) to the acrylic polymer used for the holder end-plate. The results for a 3-sphere chain are shown in Fig. 11. In common with the results for a 3-sphere chain of steel spheres of the same size, Fig. 10, only one sub-harmonic is present, and a train of solitary-wave impulses is seen. The results for tungsten carbide (Fig. 12) indicate again that only one sub-harmonic is present, although in this case of a higher amplitude than for either steel or Delrin. These results seem to indicate that changing the sphere material is less important than changing the number of spheres in a chain in terms of determining the main frequency components present in the resultant train of solitary wave impulses.

Table 2. Material properties of three materials used in these experiments.

Material	Bulk acoustic velocity c (ms^{-1})	Density ρ ($\times 10^3$ kg m^{-3})	Acoustic impedance Z (MRayls)
Acrylic polymer	2710	1.19	3.26
Chrome steel	5900	7.83	46
Delrin polymer	2400	1.42	3.45
Tungsten carbide	6656	13.8	91.9

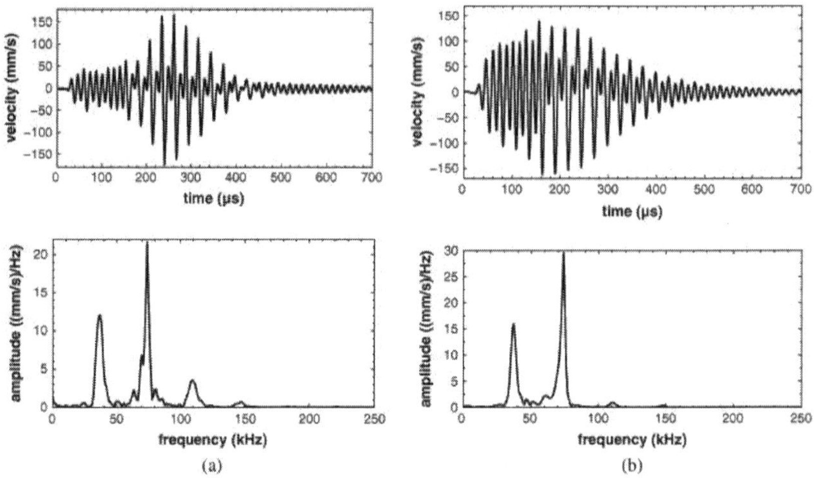

Figure. 11. Experimental result for a 3-sphere chain of Delrin spheres of 1 mm diameter, using an input waveform of 10 cycles at 73 kHz and with v_m = 634 mm s^{-1}. (a) Experimental result and (b) theoretical prediction, showing waveform (top) and frequency spectrum (bottom) in each case.

5. DISCUSSION

The above has demonstrated that interesting effects occur when high amplitude ultrasonic sinusoidal signals are input into a resonant chain of spheres. For a situation where the pre-compression F_0 is very small, the output is highly sensitive to both the exact conditions of the chain, and small changes to the input. At high input amplitudes, strongly nonlinear behaviour is observed due to Hertzian contact between spheres. Provided the amplitude and tone-burst length of the input signal are at chosen correctly, NNM resonances can exist in the form of both harmonics and sub-harmonics of the main input frequency. The effect is to create a series of solitary-wave impulses, whose characteristics depend on the structure of the chain. This type of NNM behaviour is only observed in our experiments in a certain number of chain lengths using 1 mm chrome steel spheres, namely 3, 6 and 10 spheres in the present experiments. It is thus postulated that, as we are using a fixed input frequency and sphere material, only certain chain lengths will allow resonances in the form of NNMs to be present. Modelling studies using the dynamic equations of motion presented above confirm this behaviour. Thus, the characteristics of the chain will have to be chosen carefully if solitary wave impulses are to be generated using a specific input frequency (the latter being fixed in our experiments because of the use of a highly-resonant horn as an ultrasonic source).

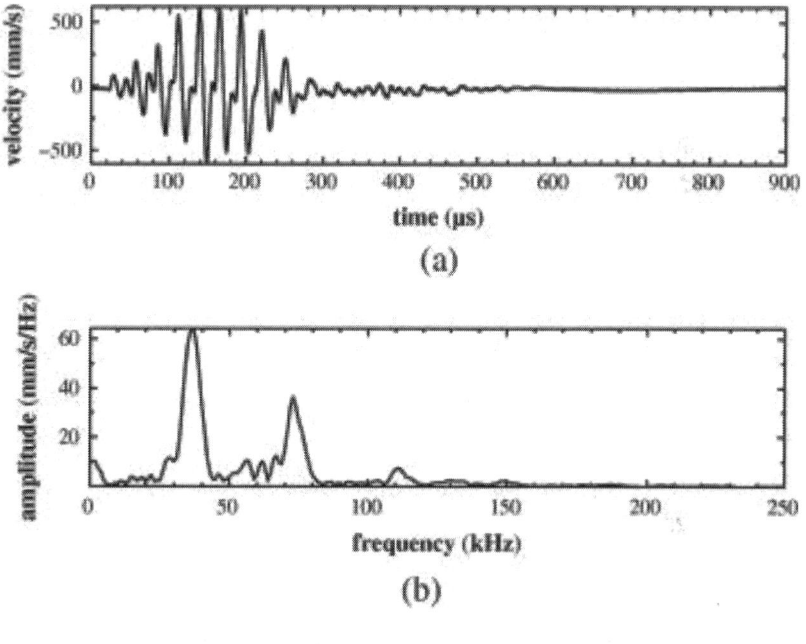

Figure. 12. Experimental result for a 3-sphere chain of tungsten carbide spheres of 1 mm diameter, using an input waveform of 10 cycles at 73 kHz and with v_m = 634 mm s^{-1}, showing (a) waveform and (b) frequency spectrum.

The experiments have shown that the output is very dependent on the input conditions. The input has to be fine-tuned to produce time-domain impulses from the sinusoidal input; both the amplitude and number of cycles in the input waveform have to be adjusted to achieve the best results for a given spherical chain with allowed NNMs. The number of spheres present at a particular frequency has a big effect on the resultant output; however, for a fixed chain length, the material properties of the spheres seems to be of secondary importance in determining the main features present in the resultant waveforms.

Modelling the behaviour by solving discrete dynamic equations of motion for each sphere in the chain is able to predict the main features of the observed experimental results. Strongly nonlinear oscillations are predicted, involving both compression and loss of contact between spheres, and these are modelled within the dynamic equations. The experimental conditions meant that some level of static pre-compression force F_0 was always present; this modified the behaviour throughout, especially at low input amplitudes, where the behaviour was weakly non-linear. At low values of F_0, however, the deliberate use of narrow bandwidth inputs of high amplitude meant that resonances in the form of NNMs were encouraged, leading to prominent harmonic and sub-harmonic generation. It is the presence of multiple NNM resonances that allows the formation of impulsive signals at high amplitude. In our experiments, the output signals can reach values of 50% of that of the input (see Fig. 9). This means that the impulses themselves are of reasonably high amplitude, leading to the possibility of uses in therapeutic ultrasound.

6. CONCLUSIONS

It has been demonstrated that impulses can be generated from high amplitude, narrow bandwidth inputs into chains of steel spheres. These are only created when the experimental conditions are correctly fine-tuned, in terms of the input waveform characteristics and the applied pre-compression force for a given chain. When this is done, a set of NNMs can be excited, which form a set of harmonics and sub-harmonics of the input frequency; it is this that leads to the creation of a train of impulses, the characteristics of which depend on the chain length for a given sphere material and diameter. These effects only occur for very small pre-compression forces F_0, where strongly non-linear behaviour occurs. These results are likely to have uses in many branches of ultrasound, where the creation of high amplitude impulses is required. This includes biomedical therapeutic ultrasound, targeted drug delivery, sonochemistry and many other fields.

ACKNOWLEDGEMENT

The authors gratefully acknowledge funding from the Engineering and Physical Sciences Research Council (UK) via grant number EP/K030159/1.

REFERENCES

1. V.F. Nesterenko, Propagation of nonlinear compression pulses in granular media, J. Appl. Mech. Tech. Phys. 24 (1983) 733–743.

2. C. Coste, E. Falcon, S. Fauve, Solitary waves in a chain of beads under Hertz contact, Phys. Rev. E 56 (1997) 6104–6117.

3. S. Sen, J. Hong, E. Avalos, R. Doney, Solitary waves in a granular chain, Phys. Rep. 462 (2008) 21–66.

4. R.S. Mackay, Solitary waves in a chain of beads under Hertz contact, Phys. Lett. A 251 (1999) 191–192.

5. V.F. Nesterenko, Dynamics of Heterogeneous Materials, Springer-Verlag, New York, 2001.

6. C. Daraio, V.F. Nesterenko, E. Herbold, S. Jin, Strongly nonlinear waves in a chain of Teflon beads, Phys. Rev. E 72 (2005) 016603.

7. E.B. Herbold, V.F. Nesterenko, Solitary and shock waves in discrete strongly nonlinear double power-law materials, Appl. Phys. Lett. 90 (2007) 261902.

8. K.R. Jayaprakash, Y. Starosvetsky, A.F. Vakakis, A new family of solitary waves in granular dimmer chains with no pre-compression, Phys. Rev. E 83 (2011) 036606.

9. C. Daraio, V.F. Nesterenko, E. Herbold, S. Jin, Tunability of solitary wave properties in one-dimensional strongly nonlinear phononic crystals, Phys. Rev. E 73 (2006) 026610.

10. A. Spadoni, C. Daraio, Generation and control of sound bullets with a nonlinear acoustic lens, Proc. Natl. Acad. Sci. 107 (2010) 7230–7234.

11. C.M. Donahue, P.W.J. Anzel, L. Bonanomi, T.A. Keller, C. Daraio, Experimental realization of a nonlinear acoustic lens with a tunable focus, Appl. Phys. Lett. 104 (2014) 014103.

12. K.R. Jayaprakash, Y. Starosvetsky, A.F. Vakakis, M. Peeters, G. Kerschen, Nonlinear normal modes and band zones in granular chains with no precompression, Nonlinear Dyn. 63 (2011) 359–385.

13. Y. Starosvetsky, A.F. Vakakis, Traveling waves and localized modes in onedimensional homogeneous granular chains with no pre-compression, Phys. Rev. E 82 (2009) 026603-1–02660314.

14. J. Lydon, K.R. Jayaprakash, D. Ngo, Y. Starosvetsky, A.F. Vakakis, C. Daraio, Frequency bands of strongly nonlinear homogeneous granular systems, Phys. Rev. E 88 (2013) 012206(9).

15. D.A. Hutchins, J. Yang, O. Akanji, P.J. Thomas, L.A.J. Davis, S. Freear, S. Harput, N. Saffari, P. Gelat, Evolution of ultrasonic impulses in chains of spheres using resonant excitation, Eur. Phys. Lett. 109 (2015) 54002.

16. N.S. Martys, R.D. Mountain, Velocity Verlet algorithm for dissipative-particledynamics-based models of suspensions, Phys. Rev. E 59 (1999) 3733–3736.

17. A. Eller, H.G. Flynn, Generation of subharmonics of order one-half by bubbles in a sound field, J. Acoust. Soc. Am. 46 (1969) 722.

18. S. Harput, M. Arif, J. McLaughlan, D.M.J. Cowell, S. Freear, The effect of amplitude modulation on subharmonic imaging with chirp excitation, IEEE Trans. Ultrason. Ferroelectr. Freq. Control 60 (2013) 2532–2544.

CHAPTER 6

Experimental Ultrasound Transmission through Fluid-Solid and Air-Solid Phononic Plates

Vicente Gómez-Lozano, Constanza Rubio *, Pilar Candelas, Antonio Uris and Francisco Belmar

Centro de Tecnologías Físicas, Universitat Politécnica de Valencia, Avd. Los Naranjos, s/n., Valencia 46022, Spain

ABSTRACT

Underwater ultrasonic transmissions for fluid-solid and air-solid phononic brass plates are reported in this work. Although the structure is roughly the same, experimental results show very different behaviour between fluid-solid and air-solid phononic plates, due to most of the properties of the fluid-solid perforated plates rely on Fabry-Perot resonances, Wood anomalies and Lamb modes. In air-solid phononic plates Fabry-Perot resonance is highly attenuated due to impedances difference between air and water, and therefore some transmission modes are now distinguishable due to surface modes coupling.

KEYWORDS

subwalength hole array; perforated plate; phononic plate; Lamb waves

1. INTRODUCTION

At the end of the 19th century Lord Rayleigh [1] studied the reflection coefficient of a one-dimensional grating. This was the first study of sound interaction with periodic structures. Since then there have been major studies of sound wave propagation through periodic structures. During the last decades, much attention has been paid to phononic crystal. The beginning of the research in the field of phononic crystals started nearly two decades ago considering mainly bulk wave propagation [2,3]. Concepts such as band-gaps, waveguiding, and negative refraction among others rapidly extended to plate guided waves (Lamb waves). Theoretical [4,5,6] and experimental [7,8,9] studies dealt firstly with band gap and waveguiding phenomena for air-solid and solid-solid phononic plates. Most of the work done so far for phononic plates restricts the sound wave propagation to the solid and the effect of the fluid (which could be treated as vacuum) is then neglected. Recently, phononic crystal strips are applied to enhance the quality factor of aluminum nitride (AlN) electroacoustic resonators utilizing the lowest symmetric (S_0) Lamb wave mode thanks to strong reflection in acoustic band gaps [10].

On the other hand, a periodically perforated plate structure has been brought into focus while researching on the acoustic counterpart of Extraordinary Optical Transmission [11,12]. Resonant full transmission [13,14], Wood anomalies [15,16], and non-leaky modes [17,18] have been reported neglecting the role of the solid and constraining the wave propagation to the fluid. These two complementary points of view can be combined via fluid-solid coupling. Although more difficult to deal with, this unified perspective enables the observation of Extraordinary Sound Screening [19], complex interactions between Lamb-like modes and hole resonances [20,21], and full transmission through non perforated (corrugated) plates [22]. It is clear that the fluid-solid coupling plays a crucial role in the acoustical properties of phononic plates, portraying a wide range of different behaviours.

In this letter we provide experimental results for fluid loaded fluid-solid and air-solid phononic plates. Angle resolved transmission measurements of underwater ultrasound reveal key aspects of the different behaviour of phononic plates having fluid or air inclusions.

2. MATERIALS AND METHODS

Underwater ultrasound has been chosen to measure the acoustic transmission through perforated plates. The geometrical parameters to describe perforated plates are the diameter of the hole d, the array periodicity a and the plate thickness h. The experimental setup is based on the well-known ultrasonic immersion transmission technique. This technique makes use of a couple of transmitter/receiver ultrasonic transducers. Two different couples of transducers were used: one couple with a center frequency of 250 kHz and a frequency range between 155–350 kHz and the other couple with center frequency of 500 kHz and a frequency range between 350 and 650 kHz. Each transducer was located at a distance larger than that of its nearfield distance (43 mm) from the plate and aligned with respect to the plate. A pulse is launched by the emitter piston transducer through the inspected plate. Then, the signal is detected by the receiving piston transducer and acquired by the pulser/receiver, post amplified and digitized by a digital PC oscilloscope (Picoscope model 3224, Cambridge, UK). Time domain data is finally analyzed after averaging 100 different measures and deleting unwanted reflections by means of a time window. The transmission spectrum is then calculated from the power spectrum of the signal normalized with the reference signal power spectrum measured without the sample plate. Typically, angle dependent measurements were done in angle steps of 1° and comprising from 0° to 60°.

The measurements were made using brass plates with 350 mm in width and 450 mm in length ($\rho = 7890$ kg/m^3, $c_l = 5670$ m/s, $c_t = 3230$ m/s) and 2 mm thickness, immersed in water ($\rho = 1000$ kg/m^3, c_l = 1480 m/s).

A brass plate drilled with periodically square distribution of circular holes having diameter, *d*, of 1 mm and a unit cell period, *a*, of 3 mm, was used (Figure 1). Starting from the perforated plate the fabrication of an air-solid phononic plate was carried out by sticking, carefully, a film (20 μm thick) around the plate, taking care that did not appear undesired air bubbles when the film was adhered.

3. RESULTS AND DISCUSSION

When an incident sound pressure wave impinges on the plate, gives rise a reflected and a transmitted sound pressure wave. In the solid, high transmission values can be observed revealing a complex dispersion. For this case, three kinds of modes can be distinguished: Scholte-Stoneley mode, symmetric leaky Lamb modes, and antisymmetric leaky Lamb modes. The Scholte-Stoneley mode propagates across the fluid-solid interface with a phase speed slightly slower than that of the water. At low frequencies, this mode is mixed with the A_0 mode and slowly converges to the sound line as ω increases. Leaky Lamb modes are guided waves produced due to the multiple reflections of longitudinal and in-plane transverse modes at both plate-fluid interfaces. Figure 2 shows the different shapes of the symmetric and antisymmetric modes in a 2 mm thickness brass plate. Although there was not periodicity due to it was not still drilled, in order to compare with subsecuent results, these are plotted in terms of the normalized frequencies $k_0 \cdot a/\pi$ and $k_\parallel \cdot a/\pi$, where *a* is the array periodicity of the perforated plate, where k_0 is the incident wavenumber and k_l is k_0 projection on the plate.

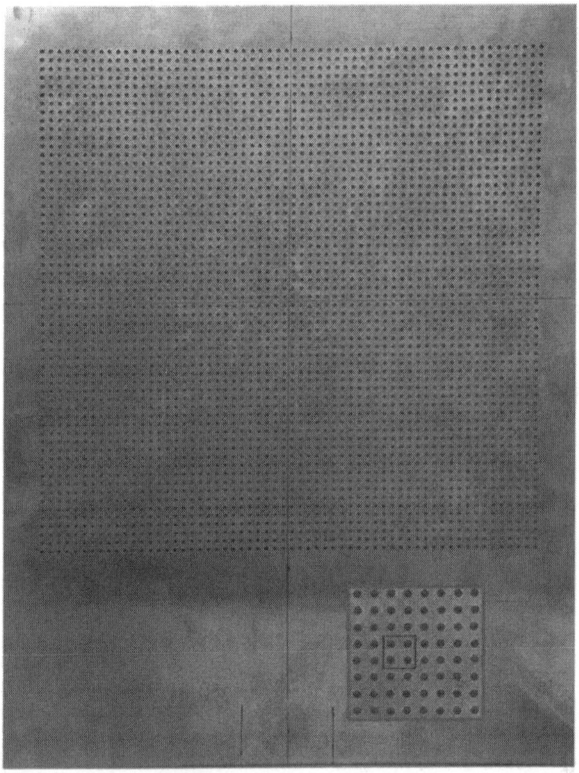

Figure 1. Structure and elementary cell for a brass plate perforated with periodically square distribution of circular holes.

However, when the plate is periodically perforated, one has to consider the elastic movement of the plate coupled with the surrounding fluid not only at the plate free surface but also inside the apertures. In order to evaluate the physical phenomena, without taking into account the elastic movement of the plate, the transmission sound power can be calculated by solving the sound wave equation within a hard-solid model [23], that is, assuming that the pressure field does not penetrate into the material plate because the water/plate impedance is considered infinite. Figure 3 shows the calculated transmitted sound power coefficient, τ, for the periodically square distribution of circular holes considered and it is observed the Fabry-Perot full transmission at $k_0 \cdot a/\pi = 1.2$ and the Wood anomaly minima at $k_0 \cdot a/\pi = 2$ that is evident when the incidence angle is varied. Due to the elastic movement of the plate had not been taken into account, Lamb and Scholte-Stoneley waves are not observed.

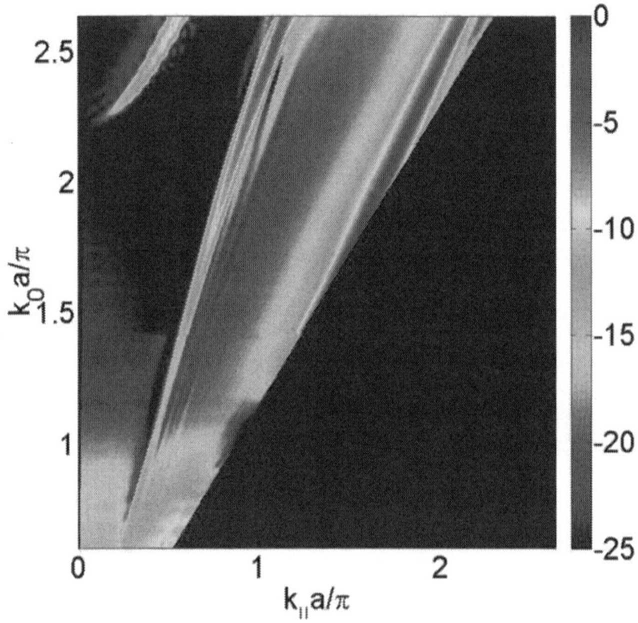

Figure 2. Measured transmitted sound power coefficient, τ, as a function of the parallel wavevector $k_{||} \cdot a/\pi$ in the ΓX direction and the normalized frequency $k_0 \cdot a/\pi$, where a the array periodicity. The transmitted sound is represented in dB scale.

Figure 4 depicts the experimental transmitted sound power τ, as a function of the parallel wavevector $k_{||} \cdot a/\pi$ in the ΓXdirection and the normalized frequency $k_0 \cdot a/2\pi$, of the brass plate perforated of thickness h = 2 mm with holes of diameter d= 1 mm arranged periodically in a square lattice of period a = 3 mm and immersed in water. Complex interaction between minima and maxima is present in the spectra and it makes clear that the symmetry of the array results in a high angular dependence of the spectra. For $k_{\|}\cdot a/\pi$ near zero (normal incidence) the Fabry-Perot full transmission at $k_0\cdot a/\pi = 1.2$ and Wood anomaly can be clearly seen both thickness at $k_0\cdot a/\pi \approx 2$. However, the modes appearing from the bottom of the figures are not predicted by the Wood anomaly and are related to leaky surface modes. These surface modes are leaky Lamb modes observed in the homogeneous plate (see Figure 2) arising from the plate vibration and the solid-fluid coupling.

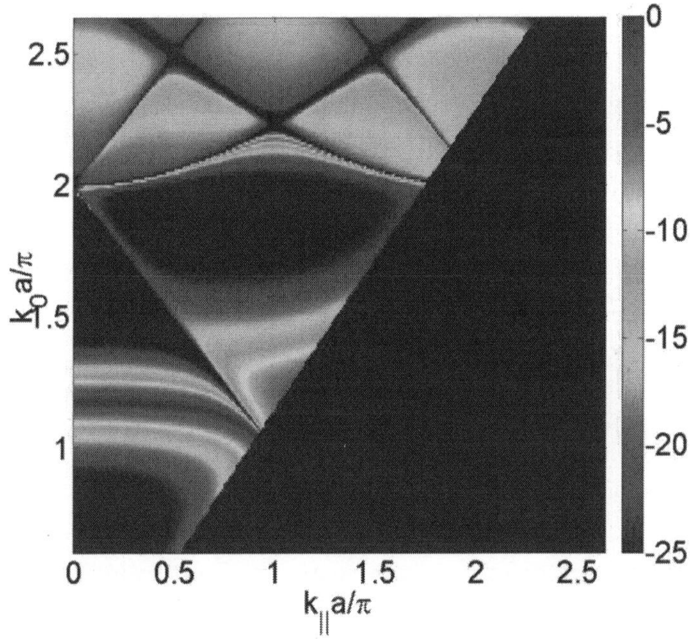

Figure 3. Calculated transmitted sound power coefficient, τ, as a function of the parallel wavevector $k_{\|}\cdot a/\pi$ in the ΓX direction and the normalized frequency $k_0\cdot a/\pi$, for the square distribution of circular holes having diameter, d, of 1 mm and a unit cell period, a, of 3 mm. The transmitted sound is represented in dB scale.

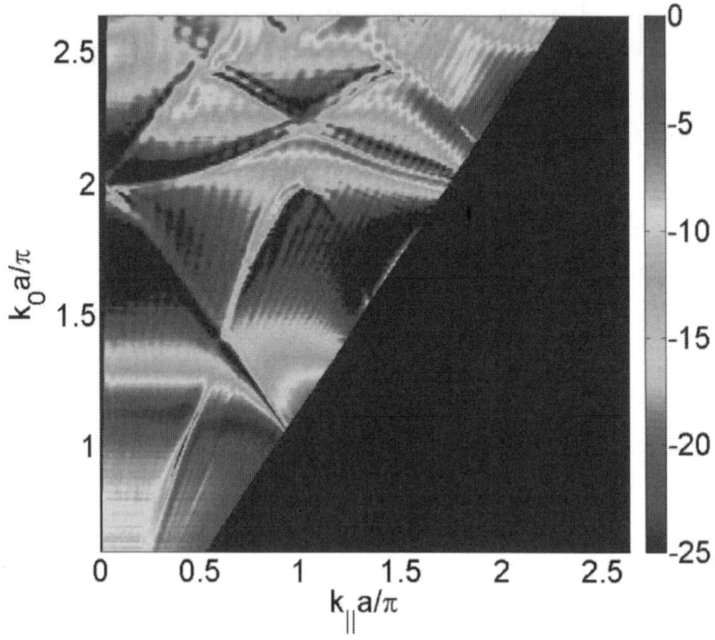

Figure 4. Measured transmitted sound power coefficient, τ, as a function of the parallel wavevector $k_{||} \cdot a/\pi$ in the Γ*X* direction and the normalized frequency $k_0 \cdot a/\pi$, for the square distribution of circular holes having diameter, *d*, of 1 mm and a unit cell period,*a*, of 3 mm. The transmitted sound is represented in dB scale. The transmitted sound is represented in dB scale.

The experimental dispersion of the same brass plate but having air filled holes is showed in Figure 5. As the wave interaction at an air-solid interface is different from that of a fluid-solid interface, the transmission properties of the phononic plate differ from the previous case. As the holes have been filled with air, the Fabry-Perot resonance was highly attenuated due to the impedance difference between water and air. Zero order Lamb modes, Scholte-Stoneley mode near grazing incidence and Wood anomalies are clearly distinguishable but there are new transmission modes that have not been detected in the case of the water filled holes.

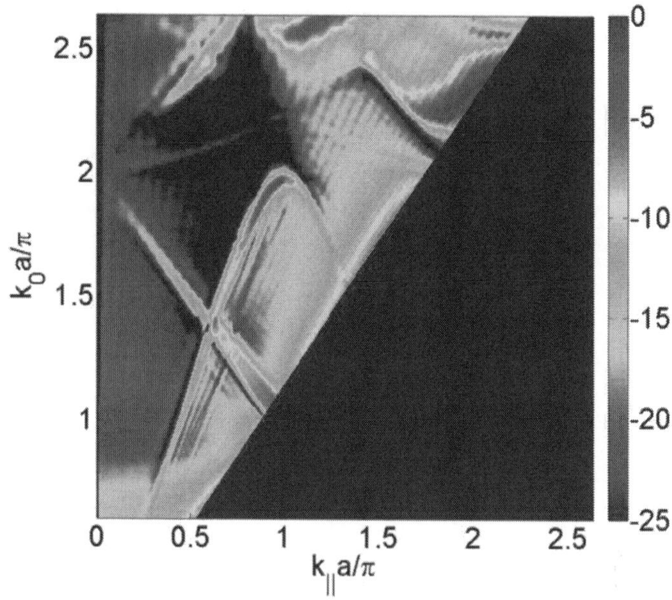

Figure 5. Measured transmitted sound power coefficient, τ, as a function of the parallel wavevector $k_{\|}\cdot a/\pi$ in the Γ*X* direction and the normalized frequency $k_0\cdot a/\pi$, for the square distribution of circular holes having diameter, *d*, of 1 mm and a unit cell period,*a*, of 3 mm and the holes filled with air (air-solid phononic plate). The transmitted sound is represented in dB scale.

The nature of these transmission modes can be demonstrated by calculating the dispersion diagram for the brass phononic plate. For this purpose we calculate the eigenfrequencies of the cell type with periodic boundary conditions of Floquet type with a variable vector $\vec{k}$ =(k,0,0) between 0 and π/2*a*. Figure 6 shows the dispersion diagram.

By superimposing the dispersion diagram with the graphics of the transmitted sound power coefficient as a function of the parallel wavevector for the perforated plates with water or air-filled holes (see Figure 7), it is observed that while some lines of the dispersion diagram clearly indicate a significant increase in transmission through plate, others do not exhibit.

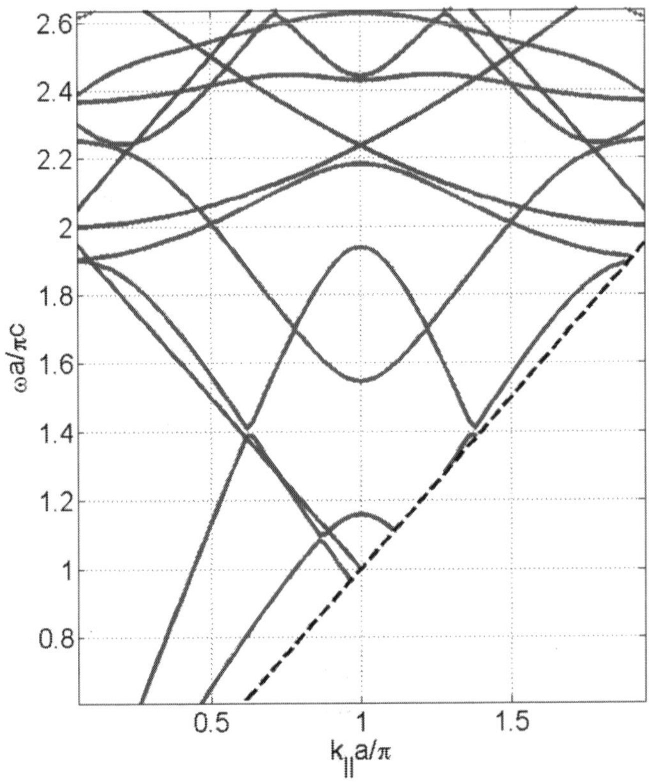

Figure 6. Calculated dispersion diagram (red lines) and Wood minima (blue lines), as a function of the parallel wavevector $k_{\|}\cdot a/\pi$ in the ΓX direction and the normalized frequency $\omega a/\pi c$, for the square distribution of circular holes having diameter, *d*, of 1 mm and a unit cell period, *a*, of 3 mm.

To analyze this fact, we calculate the displacement of the unit cell in the *z* direction for a fixed value of the parallel wavevector $k_{\|}\cdot a/\pi$ for the corresponding eigenfrequency. Figure 8 shows some eigenmodes for $k_{\|}\cdot a/\pi = 0.4$. It is noted that some are antisymmetric modes (A_1, A_2, A_3 and A_4) with respect to the forward direction of the plane wave. Clearly these eigenmodes produce poor fluid-solid coupling and lead to non-radiative bands, while symmetric eigenmodes (S_1, S_2, S_3 andS_4) produce defined transmission lines.

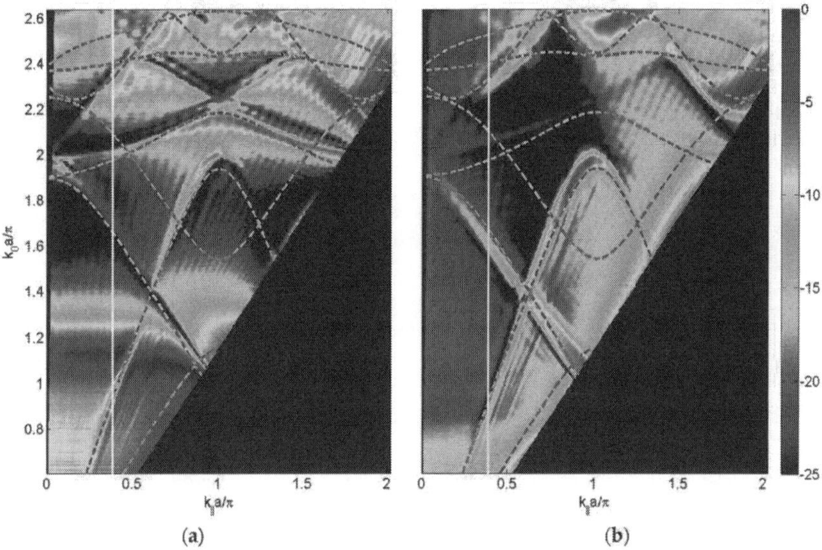

Figure 7. Dispersion diagram superimposed to measured transmitted sound power coefficient, τ, as a function of the parallel wavevector $k_{\parallel} \cdot a/\pi$ in the ΓX direction and the normalized frequency $k_0 \cdot a/\pi$, for the square distribution of circular holes having diameter, d, of 1 mm and a unit cell period, a, of 3 mm and the holes filled with (**a**) water and (**b**) air.

4. CONCLUSIONS

The results presented in this letter give a wide description of the phenomena involved in phononic plates as they also includes the coupling with the surrounding fluid. The fluid-solid phononic plates behavior is mainly governed by holes resonances and coherent diffraction effects which interact with Lamb-like modes. However, in air-solid phononic plates, zero order Lamb modes, Scholte-Stoneley mode near grazing incidence, Wood anomalies and new transmission modes are clearly distinguishable while the Fabry-Perot resonance was highly attenuated due to the impedance difference between water and air. It has been reported here that these differences are due to antisymmetric eigenmodes produce poor fluid-solid coupling and lead to non-radiative modes, while symmetric eigenmodes has high fluid-solid coupling and lead radiative modes.

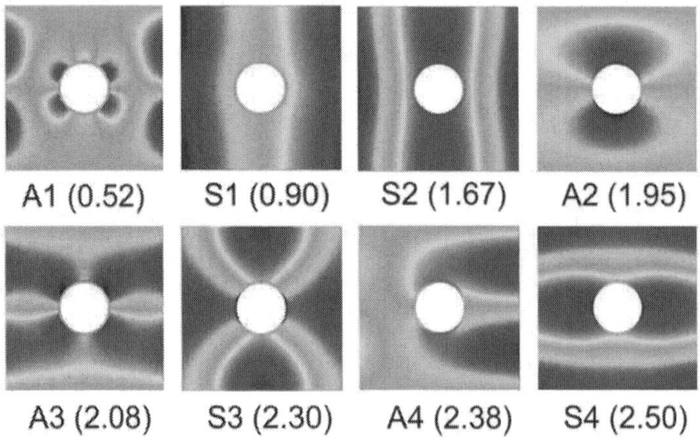

Figure 8. z-displacement of the brass unit cell for a fixed value of $k_{\parallel}\cdot a/\pi = 0.4$ The eigenmodes are labeled as S (symmetric) or A (antisymmetric) according to they behaviour to the horizontal plane wave incoming from the left side of the plots. The eigenmodes normalized frequency $k_0\cdot a/\pi$ values are despited in parenthesis. The displacement is represented in linear scale.

ACKNOWLEDGMENTS

This work has been supported by Spanish MINECO (TEC2015-70939-R) and Generalitat Valenciana (AICO/2015/119).

AUTHOR CONTRIBUTIONS

Constanza Rubio and Antonio Uris has coordinated the theoretical and experimental developments, participating in the establishment of the theory principles used in this work, as well as in the drafting of the manuscript. Vicente Gomez-Lozano has developed part of the theory used and has designed some characterization experiments and has developed some computing task, as well as has participated in the drafting of the manuscript. Pilar Candelas, Francisco Belmar have designed and carried out the characterization of the experimental setup and in computing algorithms for the treatment of data, also have participated in the analysis of the state of art and in the design of some experiments, as well as in the drafting of the manuscript.

CONFLICTS OF INTEREST

The authors declare no conflict of interest. The founding sponsors had no role in the design of the study; in the collection, analyses, or interpretation of data; in the writing of the manuscript, and in the decision to publish the results.

REFERENCES

1. Rayleigh, L. *The Theory of Sound*, 2nd ed.; Courier Dover Publications: New York, NY, USA, 1945; Volume 2, p. 89.

2. Miyashita, T. Sonic crystals and sonic waveguides. *Meas. Sci. Technol.* **2005**, *16*, R47–R63.

3. Sigalas, M.; Kushwaha, M.S.; Economou, E.N.; Kafesaki, M.; Psarobas, I.E.; Steurer, W. Classical vibrational modes in photonic lattices: Theory and experiment. *Z. Kristallogr.* **2005**, *220*, 765–809.

4. Chen, J.-J.; Qin, B.; Cheng, J.C. Complete band gaps for Lamb waves in cubic thin plates with periodically placed inclusions. *Chin. Phys. Lett.* **2005**, *22*, 1706–1708.

5. Khelif, A.; Aoubiza, B.; Mohammadi, S.; Adibi, A.; Laude, V. Complete band gaps in two-dimensional phononic crystal slabs. *Phys. Rev. E* **2006**, *74*, 045510.

6. Vasseur, J.O.; Deymier, P.A.; Djafari-Rouhani, B.; Pennec, Y.; Hladky-Hennion, A.C. Absolute forbidden bands and waveguiding in two-dimensional phononic crystal plates. *Phys. Rev. B* **2008**, *77*, 085415.

7. Zhang, X.; Jackson, T.; Lafond, E.; Deymier, P.; Vasseur, J. Evidence of surface acoustic wave band gaps in the phononic crystals created on thin plates. *Appl. Phys. Lett.* **2006**, *88*, 041911.

8. Brunet, T.; Vasseur, J.; Bonello, B.; Djafari-Rouhani, B.; Hladky-Hennion, A.C. Lamb waves in phononic crystal slabs with square or rectangular symmetries. *J. Appl. Phys.* **2008**, *104*, 043506.

9. Mohammadi, S.; Eftekhar, A.A.; Khelif, A.; Hunt, W.D.; Adibi, A. Evidence of large high frequency complete phononic band gaps in silicon phononic crystal plates. *Appl. Phys. Lett.* **2008**, *92*, 221905.

10. Lin, C.-M.; Hsu, J.-C.; Senesky, D.G.; Pisano, A.P. Anchor loss reduction in AlN Lamb wave resonators using phononic crystal strip tethers. *Proc. IEEE Int. Freq. Control Symp.* **2014**.

11. Ebbesen, T.W.; Lezec, H.J.; Ghaemi, H.F.; Thio, T.; Wolf, P.A. Extraordinary optical transmission through subwavelength hole arrays. *Nature* **1998**, *391*, 667–669.

12. Garcia-Vidal, F.J.; Martin-Moreno, L.; Ebbesen, T.W.; Kuipers, L. Light passing through subwavelength apertures. *Rev. Mod. Phys.* **2010**, *82*, 729–787.

13. Zhou, L.; Kriegmann, G.A. Complete transmission through a periodically perforated rigid slab. *J. Acoust. Soc. Am.***2007**, *121*, 3288–3299.

14. Lu, M.; Liu, X.; Feng, L.; Li, J.; Huang, C.; Chen, Y.; Zhu, Y.; Zhu, S.; Ming, N. Extraordinary acoustic transmission through a 1D grating with very narrow apertures. *Phys. Rev. Lett.* **2007**, *99*, 174301.

15. Hou, B.; Mei, J.; Ke, M.; Wen, W.; Liu, Z.; Shi, J.; Sheng, P. Tuning Fabry-Perot resonances via diffraction evanescent waves. *Phys. Rev. B* **2007**, *76*, 054303.

16. Wood, R.W. On a remarkable case of uneven light in a diffraction grating. *Philos. Mag.* **1902**, *4*, 396–402.

17. Christensen, J.; Martin-Moreno, L.; Garcia-Vidal, F.J. Theory of resonant acoustic transmission subwavelength apertures. *Phys. Rev. Lett.* **2008**, *101*, 014301.

18. He, Z.; Jia, H.; Qiu, C.; Ye, Y.; Hao, R.; Ke, M.; Liu, Z. Nonleaky surface acoustic waves on a textured rigid surface.*Phys. Rev. B* **2011**, *83*, 132101.

19. Estrada, H.; Candelas, P.; Uris, A.; Belmar, F.; Garcia de Abajo, F.J.; Meseguer, F. Extraordinary sound screening in perforated plates. *Phys. Rev. Lett.* **2008**, *101*, 084302.

20. Estrada, H.; Candelas, P.; Uris, A.; Belmar, F.; Garcia de Abajo, F.J.; Meseguer, F. Influence of lattice symmetry on ultrasound transmission through plates with subwavelength aperture arrays. *Appl. Phys. Lett.* **2009**, *95*, 051906.

21. Estrada, H.; Garcia de Abajo, F.J.; Candelas, P.; Uris, A.; Belmar, F.; Meseguer, F. Angle-dependent ultrasonic transmission through plates with subwavelength hole arrays. *Phys. Rev. Lett.* **2009**, *102*, 144301.

22. He, Z.; Jia, H.; Qiu, C.; Peng, S.; Mei, X.; Cai, F.; Peng, P.; Ke, M.; Liu, Z. Acoustic transmission enhancement through a periodically structured stiff plate without any opening. *Phys. Rev. Lett.* **2010**, *105*, 074301.

23. Estrada, H.; Candelas, P.; Uris, A.; Belmar, F.; Meseguer, F.; Garcia de Abajo, F.J. Sound transmission through perforated plates with subwavelength hole arrays: A rigid-solid model. *Wave Motion* 2011, *48*, 235–242.

CHAPTER 7

A role of ultrasonic frequency and power on oil mobilization in underground petroleum reservoirs

Hossein Hamidi, Roozbeh Rafati, Radzuan Bin Junin, Mohammad A. Manan

ABSTRACT

In oil industry, the reduction of oil production is of major concern as world's necessity for oil increases. Therefore, developing and applying new techniques to mobilize residual oil left in the reservoir is important. Ultrasound technique is one of the unconventional methods to increase the productivity of oil wells, but in spite of many laboratory experiments on oil mobilization under ultrasound in porous media the precise mechanisms are weakly understood. Therefore, it is vital to perform basic experiments to achieve a good knowledge and deep insight into the mechanisms. To recognize the reason why ultrasound is able to mobilize residual oil left in pores, one must remember why oil droplet is trapped. Residual oil is left in pores because of insisting capillary forces. So, in this paper a mechanism related to capillary forces are developed to clarify the effect of ultrasound on mobilization of residual oil in porous media. In addition, some experiments are conducted with a 2D glass micro-model in which the mobilization of oil is observed using a digital microscope and camera. It was concluded from the result of experiments that the oil mobilization was proportional to ultrasound power and frequency.

KEYWORDS

Ultrasonic, Power, Frequency, Mobilization, Oil

1. INTRODUCTION

After primary and secondary recovery of an oil field, using some conventional-enhanced oil recovery methods such as chemical flooding (surfactant flooding), polymer flooding, gas flooding, steam flooding for improving oil recovery are definitely required. However, these methods are expensive to use, need a wide range of a surface apparatus, generate dangerous environmental results, and have technical limitations. Therefore, some unconventional methods have been recommended with lower application risk. One of these unconventional methods is the application of ultrasound for oil reservoir stimulation. To clarify the effects of ultrasonic waves on oil recovery many laboratory experiments and field observations have been conducted during the past 50 years.

The interest in recovery of oil from oil reservoirs under vibration dates back to 1950s when increased oil production was observed because of cultural noise and earthquakes. The viability of this method to increase oil recovery was first shown by Duhon and Campbell (1965). Chen (1969) and Fairbanks and Chen (1971) showed the effect of ultrasonic waves with the frequency of about 20 kHz and power of 10 W/cm^2 on the oil and water flow through porous sandstone and through a capillary. They concluded that, ultrasound enhanced the percolation rate in the case of a porous medium. They also showed that this increase is only slightly due to ultrasonic heating.

Simkin and Surguchev (1991) and Simkin (1993) pointed out that oil droplet size was increased after applying ultrasonic waves. This increase was attributed to coalescence of oil droplets because of Bjerknes forces under ultrasound.

Kostrov and Wooden (2001) presented some possible oil recovery mechanisms as the following: (1) dislodging the oil droplet from the pore wall into the flow of fluid, (2) coalescence of two or more the oil droplets (oil films surrounding the formation grains) into the big one having higher mobility. Their results showed an increase in oil production to about 40% of total oil production.

Amro et al. (2007) conducted some experiments on core samples to show the effect of ultrasonic waves on oil mobilization. They concluded that oil displacement increased after applying ultrasonic waves and relative permeability changed. Furthermore, they concluded that applying ultrasound at residual oil saturation led to higher oil recovery compared to original oil in place.

Beresnev et al. (2005) explained a capillary-related mechanism to describe the mobilization of oil droplet inside porous media after applying ultrasonic waves. They conducted this study at pore scale on a non-wet oil droplet which was trapped at the entry of a pore in a channel, as showed in Fig. 1. As seen, the radius of the left meniscus is greater than the right one ($R_{\text{left}} > R_{\text{right}}$). Therefore, an inequality of capillary pressure (ΔP) is generated inside the oil droplet that resisting the external pressure. As the oil droplet move more forward to the throat, the inequality of capillary pressure increases more. When the oil droplet is trapped, the pressure gradient (ΔP) inside it equals to the external pressure gradient (ΔP_s). To move the droplets through the throat, ΔP_s must be increased (more than ΔP). In this case, the droplet starts to move. This additional pressure can be created by ultrasonic waves (P_U). It means that when $\Delta P_s + P_U > \Delta P$ then the oil droplet can pass through the throat. In addition, the force generated by ultrasonic wave increases with the increase of ultrasound power.

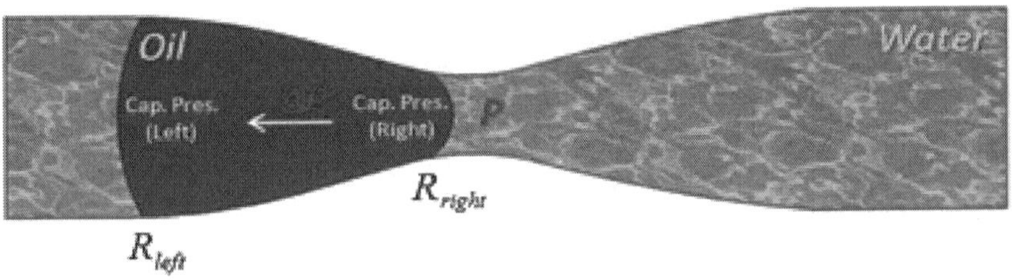

Figure. 1 A trapped oil under a ΔP (pressure gradient). The droplet radii are shown as R_{right} andR_{left}. The term, *Cap. Pres.* refers to capillary pressure

Based on the above literature, it can be predicted that the oil mobilization enhances with increasing ultrasound power. So, in this study to prove the claim, a comprehensive series of macro-model (Core sample) and micro-model experiments were done to show the effect of ultrasonic waves on oil recovery rate and oil mobilization for different medium wettability (oil wet and water wet), and oil viscosities (15, 100, and 400 cp). In addition, the main parameters of ultrasound sources such as frequency (25 and 40 kHz) and power (1–500 W) were altered.

2. EXPERIMENTAL SETUP AND PROCEDURE

2.1 Macro-model experiments

For macro-model experiments, co-current (all sides open to flow) Berea sandstone cores were used with the dimension of 6 × 3 cm ($L \times D$). After saturating of the cores by oil in the vacuum, they were put inside of an imbibition cell filled up with brine. Then, for ultrasonic tests, the cells were placed inside an ultrasonic bath (ultrasonic cleaner) filled up with de-aerated water. We used two ultrasonic generators to produce the vibrational energy at frequencies of 25 and 40 kHz and power of 250 and 500 W, respectively. The schematic diagram of the imbibition experiment apparatus is shown in Fig. 2. The mineral oils used in experiments had various viscosities like 15, 100, and 400 cp. In some of the experiments for investigating the effect of different ultrasonic frequencies on oil recovery, we used the minimum energy output (1 W). The experimental runs are shown in Table 1.

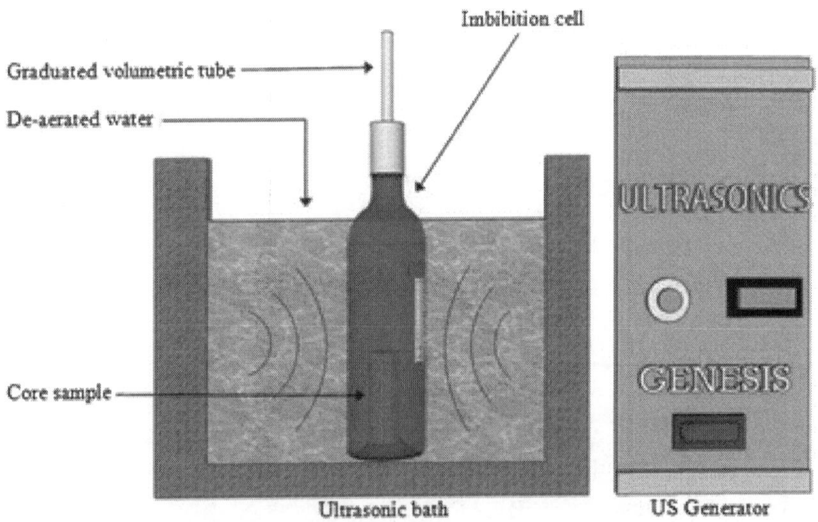

Figure. 2 Experimental setup for imbibition experiments

2.2 Micro-model experiments

According to Fig. 3, after preparation and fabrication of the 2D glass micro-model, they were connected to the valves to which control the flow in and out of the model. The dimension of micro-model was about 60 × 60 mm (Fig. 4). The pore and throat diameters were 740 and 280 μm, respectively. The unit was immersed into a bath which was filled with water to make a suitable surrounding for propagation of ultrasonic waves.

Table 1 Experimental runs performed in this study

No.	Wettability	Viscosity (cp)	Power (W)	Frequency (kHz)
1	Water wet	400	500	25
2	Water wet	400	250	25
3	Water wet	400	0	25
4	Water wet	100	500	25
5	Water wet	100	250	25
6	Water wet	100	0	25
7	Water wet	15	500	25
8	Water wet	15	250	25
9	Water wet	15	0	25
10	Oil wet	100	250	25
11	Oil wet	100	0	25
12	Oil wet	15	250	25
13	Oil wet	15	0	25
14	Water wet	100	1	40
15	Water wet	100	1	25
16	Water wet	15	1	40
17	Water wet	15	1	25

To run tests, at first, brine (dyed deep blue) with salinity concentration of NaCl = 0.35 wt(%), $MgCl_2$ = 0.25 wt(%), and $CaCl_2$ = 0.10 wt(%) was injected by a KD Scientific syringe pump into the micro-model to saturate the micro-model. Then, for several hours, micro-model was left intact. Afterward, the oil (dyed red) was injected at 1.0 ml/h to move the initial brine. The oil properties are shown in Table 2. The injection was continued until reaching a point in which only the oil was produced at the port. This point could be recognized using microscope and camera at the production port and throughout the micro-model. After recognizing this point the oil injection could be finished. In this stage, a condition of connate water saturation was provided throughout the micro-model. Then brine (dyed pale blue) with the salinity concentration equal to surfactant solution salinity concentration was injected into the micro-models at the injection rate of 0.018 ml/h. It must be noted that, during the injection of brine, ultrasonic waves were also applied. Ultrasonic generator that was placed at the top of the micro-model generated these ultrasonic waves. A stereo microscope, Carl Zeiss Brand, Model Stemi SV 11 was focused at the pore scale. The camera mounted on this microscope monitored the displacement of the oil by water and sent the images to a PC for analyses. For calculating pores and throats area that occupied by the oil, the Image Analyzer Software was used. In these series of experiments, the ultrasound power and frequency for each experiment were different. The experimental runs are shown in Table 3.

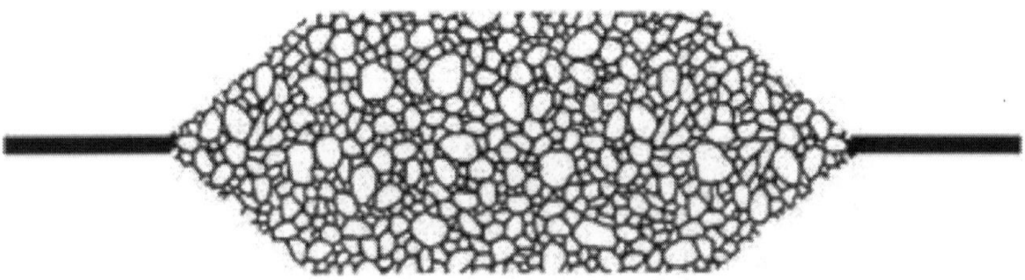

Figure. 3 Schematic of the micro-model

3. RESULTS

3.1 Macro-model experiments

The recovery curves for water-wet experiment Nos. 1–3 with oil viscosity of 400 cp are shown in Fig. 5. It is obvious that there is not any significant change in oil recovery rates due to ultrasonic waves. Therefore, the effect of ultrasonic waves on oil recovery for the experiments with heavy oil (400 cp) is trivial. Experiment Nos. 1 and 2 were performed again for repeatability. Figures 6 and 7 shows the error bars for these two experiments. Error bars showing the data variability and are applied on graphs in a measurement to show the error, or uncertainty. They are used to show that how precise a measurement is, or on the other hand, how far from the reported value the true (error free) value might be. As seen in Figs. 6 and 7, there is an early variation in oil recovery but ultimate recovery is approximately the same for the cases. The small variations at the beginning of the experiments could be ascribed to the alterations in the properties of cores since various cores were used in the experiments from the same geological block.

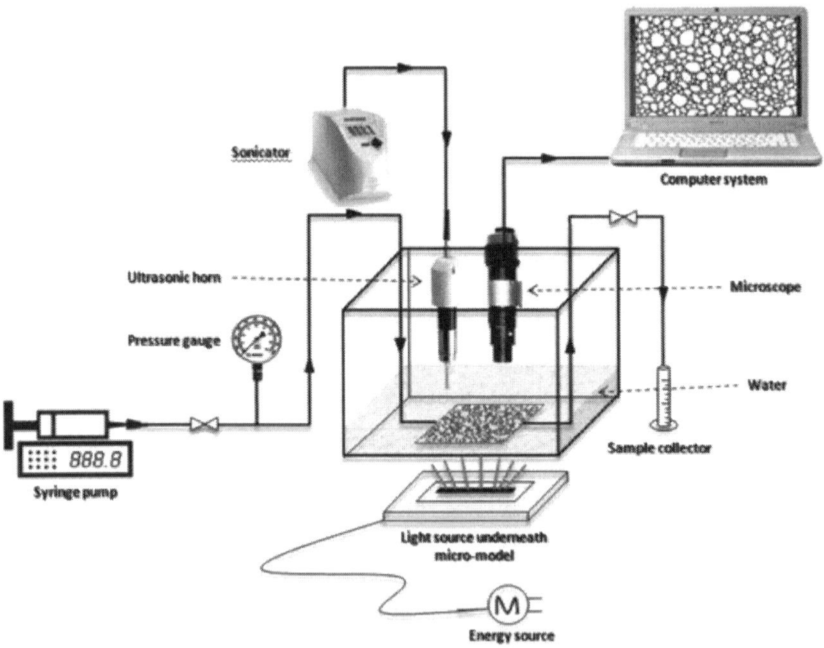

Figure. 4 Micro-model experimental setup

Table 2 Properties of the oil

T (°C)	Density (g/cm^3)	Viscosity (cp)
20	1.45	0.54

Table 3 Experimental runs performed in this study

No.	Wettability	Power (W)	Frequency (kHz)
1	Water wet	350	20
2	Water wet	350	40
3	Water wet	350	68
4	Water wet	50	40
5	Water wet	180	40
6	Water wet	350	40
7	Water wet	500	40

The recovery curves for water-wet experiment Nos. 4–6 with oil viscosity of 100 cp are shown in Fig. 8. It was observed that the ultimate oil recovery increased with increasing the power of the ultrasonic generator. Apparently, there are some possible mechanisms of oil recovery except the one explained in "Introduction" part control the process in these water-wet experiments, as following: (a) changing in interfacial tension (IFT) and wettability that were also mentioned before by Hamida and Babadagli (2007), (b) coalescence of oil droplets by Bjerknes forces that was also observed by Bjerknes (1906), (c) peristaltic movement of brine after applying ultrasound, which was mentioned before by Ganiev et al. (1989). The recovery curves for water-wet experiment Nos. 7–9 with oil viscosity of 15 cp are shown in Fig. 9. As seen, the ultimate oil recovery increased under ultrasound, especially for the tests with power of 500 W. The recovery curves for oil-wet experiment Nos. 10–13 with oil viscosity of 100 and 15 cp are shown in Fig. 10. As seen, the ultrasound has a very considerable effect on oil mobilization and oil recovery rate. In these series of experiments, the ultimate recovery increased more than 100% by ultrasound compared to water-wet cases. The oil recovery mechanisms that may be responsible in these experiments except the one described in Introduction part are IFT changes and peristaltic movement of brine inside the porous media. In oil-wet cases, aqueous phase has more affinity to flow and lesser tendencies to adhere to wall surface of pores or throats. Therefore, peristaltic movement of aqueous phase may smooth the progress of movement of oil through porous media. The coalescence of oil droplet by Bjerknes forces is not an effective mechanism for oil-wet cases because in oil-wet cases the ganglion form of oil droplets does not exist. In addition, it can be seen that the effect of ultrasonic waves on oil recovery is more significant in oil-wet cases with viscosities of 15 cp.

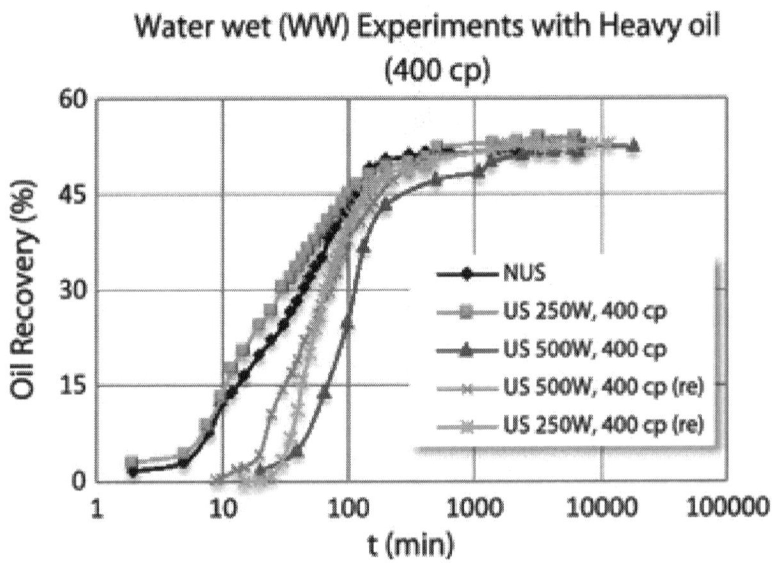

Figure. 5 Experiments 1–3, heavy oil (400 cp) with ultrasound frequency of 25 kHz

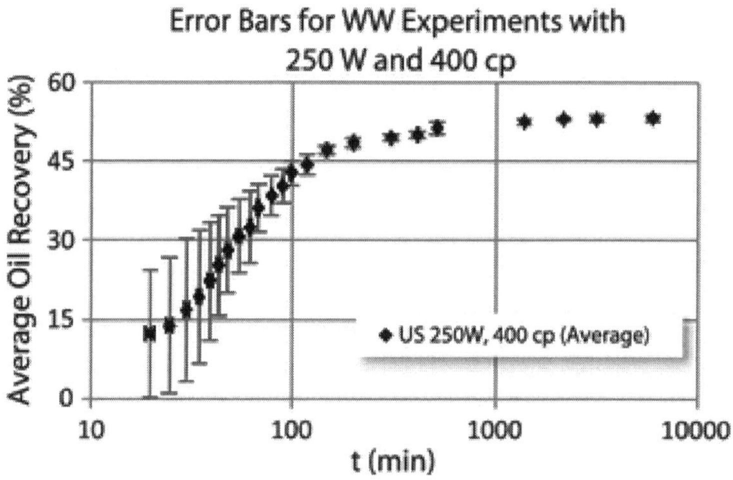

Figure. 6 *Error bars* for experiment No. 2, heavy oil (400 cp) with ultrasound power 250 kHz

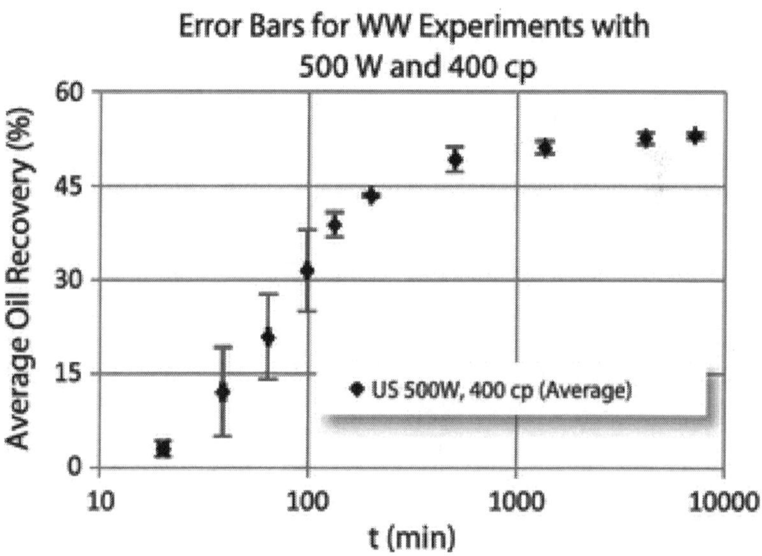

Figure. 7 *Error bars* for experiment No. 1, heavy oil (400 cp) with ultrasound power of 500 W

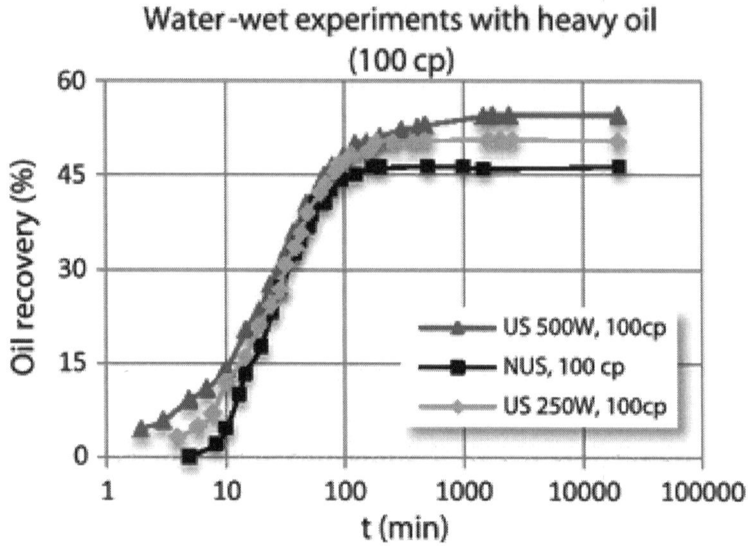

Figure. 8 Experiments 4–6, heavy oil (100 cp) with ultrasound frequency of 25 kHz

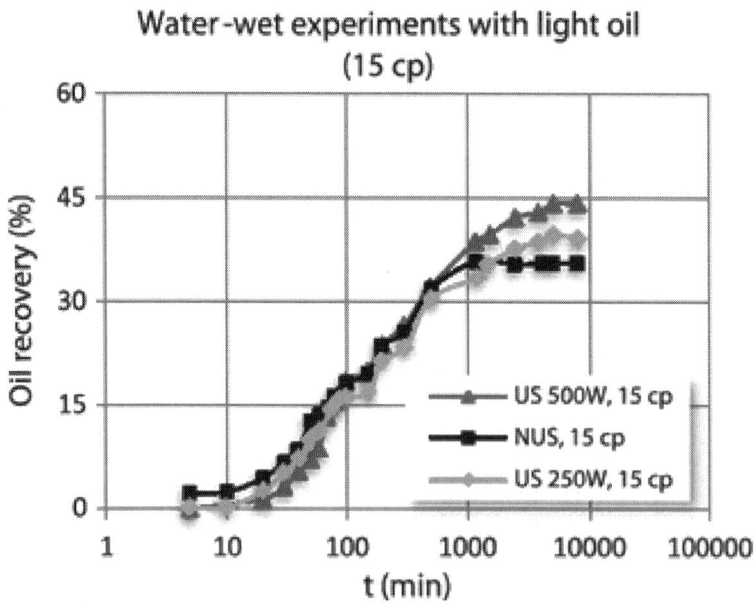

Figure. 9 Experiments 7–9, light oil (15 cp) with ultrasound frequency of 25 kHz

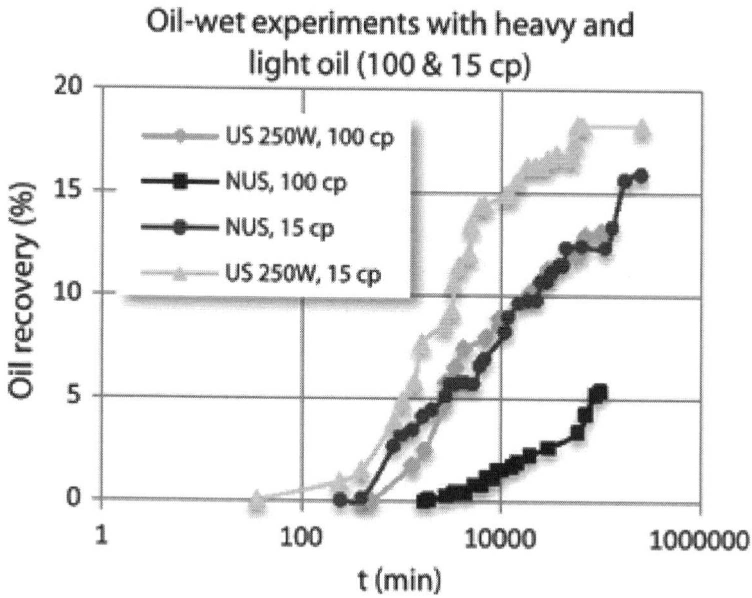

Figure. 10 Experiments 10–13, heavy and light oil (100 and 15 cp) with ultrasound frequency of 25 kHz

The recovery curves for water-wet experiment Nos. 14–17 with oil viscosity of 100 and 15 cp, low ultrasound power (1 W) and ultrasound frequencies of 25 and 40 kHz are shown in Fig. 11. These experiments were conducted to show the effect of frequency on oil recovery. The results show that for 40 kHz frequencies the recovery rate increases but ultimate recovery for all four tests (40 and 25 kHz) was the same. For repeatability, the experiment Nos. 15 and 17 were repeated. These experiments are shown by (re) in the Fig. 11. Figures 12 and 13 show the error bars for these two experiments. The repeated test No. 17 demonstrates almost the same recovery trend; however, its final recovery shows a few percent alteration compare to the first experiment No. 17. This alteration could be due to a variation in core properties or conditions of the experiments. For experiment No. 15, the repeated curve of recovery has a close match with its first one.

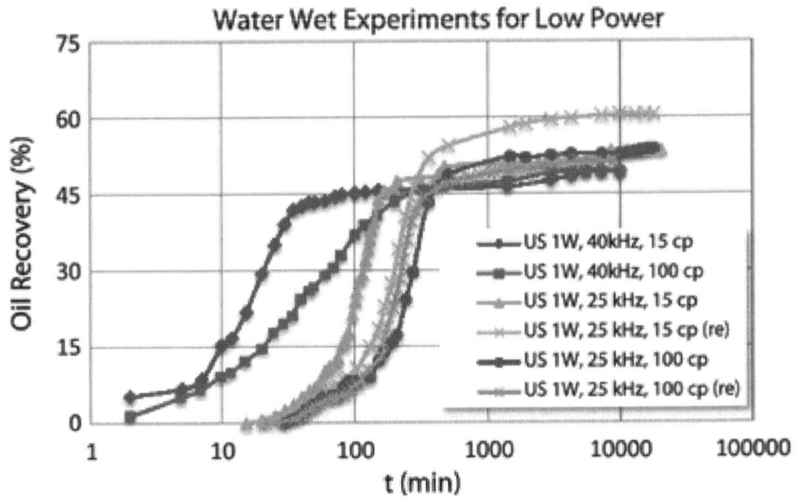

Figure. 11 Experiments 14–17, heavy and light oil (100 and 15 cp) with ultrasound frequency of 25 and 40 kHz

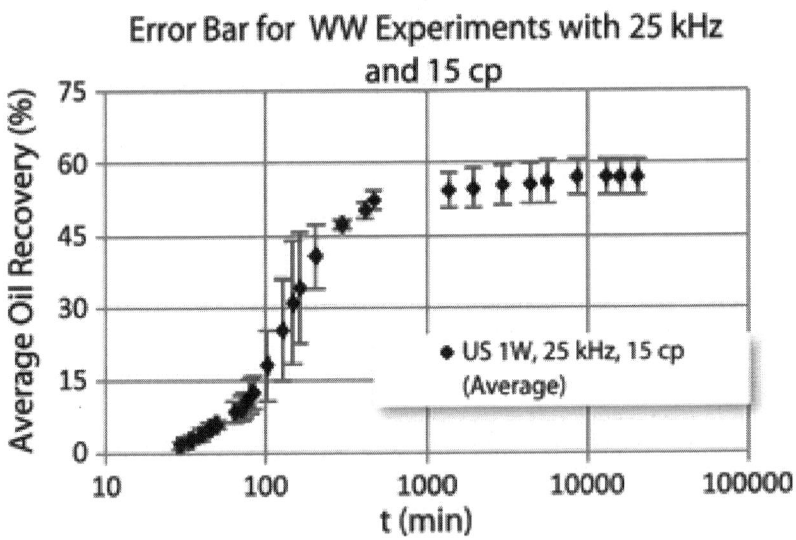

Figure. 12 *Error bars* for experiment No. 17, light oil

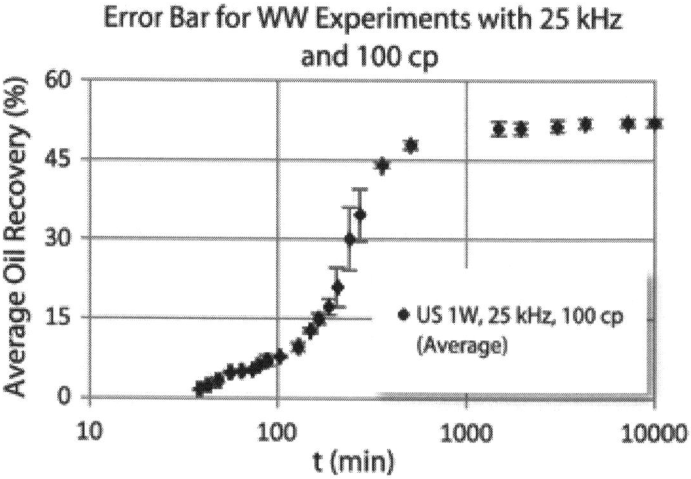

Figure. 13 *Error bars* for experiment No. 15, (15 cp) with ultrasound power of 1 W, heavy oil (100 cp) with ultrasound power of 1 W

3.2 Micro-model experiments

The result of water-wet experiments 1–3 at an ultrasound power of 350 W and different frequencies of 20, 40, and 68 kHz are shown in Fig. 14. This figure shows the residual oil saturation as a function of time. The purpose of doing these experiments was to understand the effect of ultrasonic frequency on oil mobilization. As seen, it is obvious that the oil mobilization in porous media is dependant to ultrasound frequency and residual oil decreased much faster than for the case without vibration (NUS).

For the repeatability, three experiments of water flood were conducted in the absence of ultrasound (Fig. 15). As seen, the general trends for all the three experiments are analogous, particularly at the first displacement period but they deviate as they come up to the end of process. Figure 16 shows the error bars for the experiments. These deviations are possibly due to statistical variations related to the size of the lattice. Specifically, the preliminary pathways of percolation control the saturation of the residual oil. In addition to these experiments, three other experiments were performed and repeated at ultrasound frequency of 68 kHz and power of 350 W (experiment No. 3) (Fig. 17). Figure 18 shows the error bars for the experiments. As seen, the repeatability for the tests is very good and much better than obtained for the case of NUS. Because the force generated by ultrasound exerted on all sections of the micro-model at the same time and without inequity, the probability of growing pathways of percolation which bypass great flow cell portions (as a result of that causes larger variations from one experiment to another) is reduced.

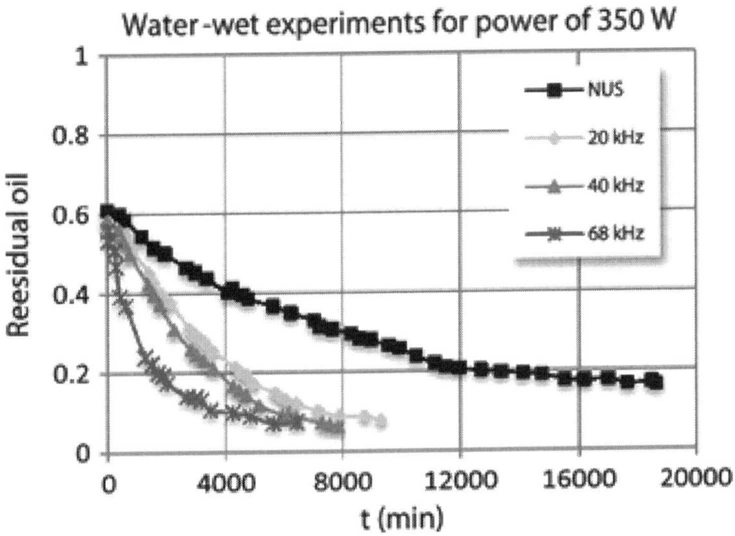

Figure. 14 Residual oil versus time for experiments done at different ultrasound frequencies and constant power

The result of water-wet experiments 4–7 at an ultrasound frequency of 40 kHz and different powers of 50, 180, 350 and 500 W are shown in Fig. 19. This figure shows the residual oil saturation as a function of time. As seen, the mobilization of oil inside porous media is also dependant to ultrasound power. Therefore, for greater powers the residual oil saturation decreased much faster than for the case without vibration (NUS). With respect to Figs. 14 and 19, it was concluded that the predictions summarized in the "Introduction" part, are validated by conducting these series of experiments.

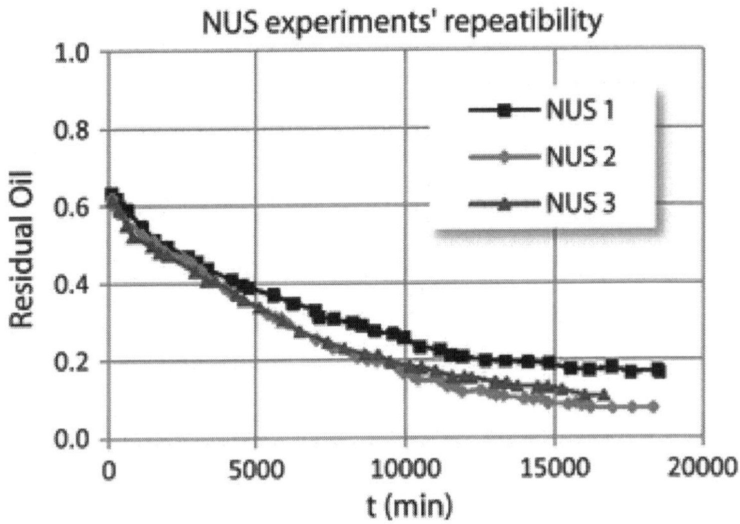

Figure. 15 Residual oil versus time for three NUS repeated experiments

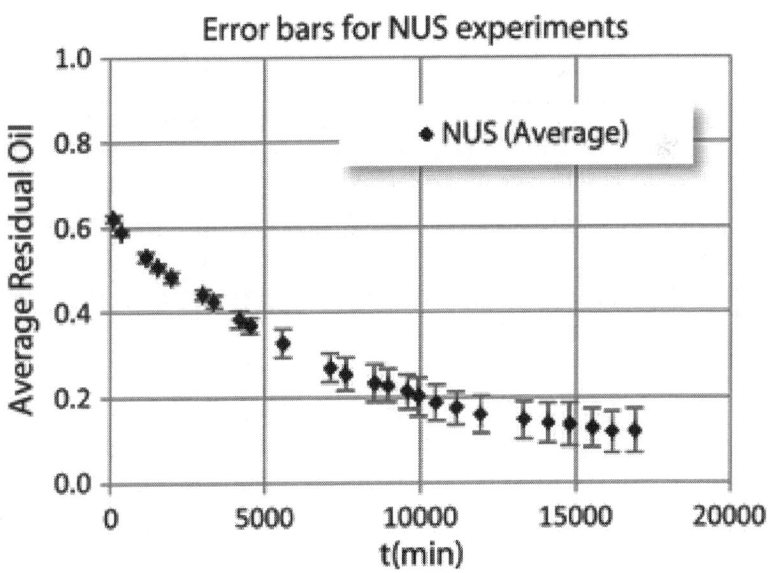

Figure. 16 *Error bars* for NUS experiments

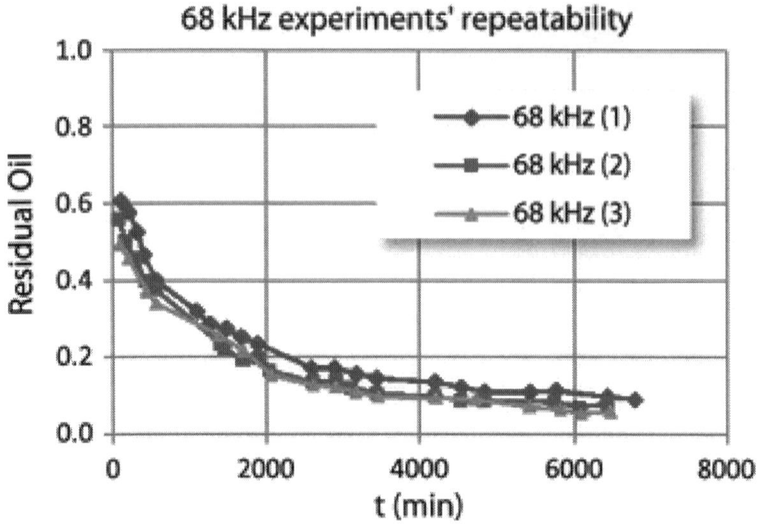

Figure. 17 Residual oil versus time for repeated experiments done with ultrasound frequency of 68 kHz and power of 350 W

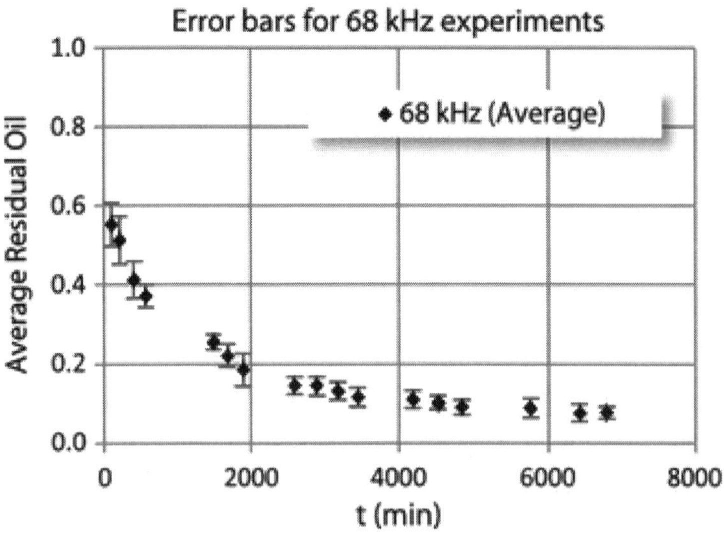

Figure. 18 *Error bars* for experiments No. 3 with experiments done with ultrasound frequency of ultrasound frequency of 68 kHz and power of 350 W

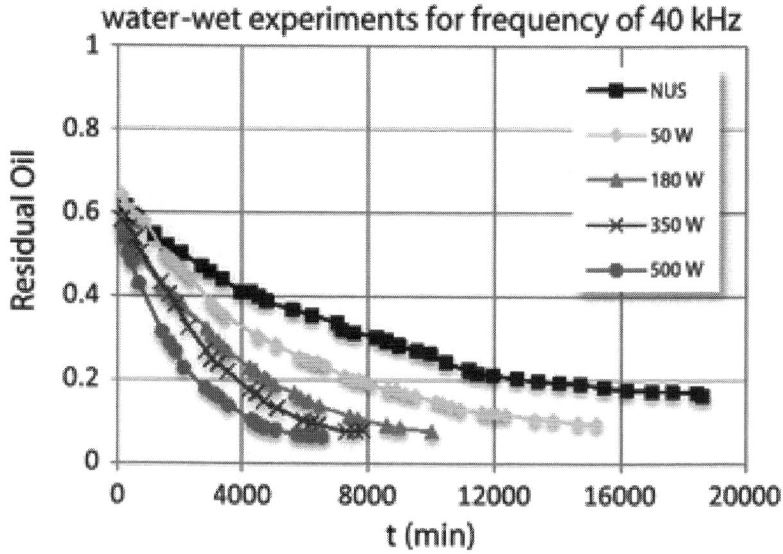

Figure. 19 Residual oil versus time for experiments done at different ultrasound power and constant frequency

4. CONCLUSIONS

In this study, the macro-model and 2D glass micro-model experiments were performed to prove the predictions of a mechanism of oil mobilization in porous media under ultrasonic waves. The result of the macro-model experiments showed that:

- the oil recovery increases under ultrasonic waves.
- The effect of ultrasound on oil recovery in a medium with high oil viscosity is less than the cases with lower viscosities.
- The effect of ultrasonic waves on oil recovery is more significant in oil-wet cases especially with lower viscosities.
- The recovery rate for higher frequency is more significant than lower frequency. Nevertheless, the ultimate recovery was almost the same for two cases.

The result of the micro-model experiments showed that:

- when the frequency increases from 20 to 68 kHz for constant ultrasound power, the oil droplet move faster in the water flood experiments.
- When the ultrasound power increases from 50 to 500 W for constant ultrasound frequency (40 kHz), the oil droplet move faster in the water flood experiments. So, the residual oil saturation decreases.

In general, the findings of this study show that the rate of oil production is proportional to the ultrasound power and frequency.

ACKNOWLEDGMENTS

The authors appreciate the Faculty of Petroleum and Renewable Energy Engineering at University Technology Malaysia in Johor Bahru, Malaysia for providing laboratory facilities in completing this work.

REFERENCES

1. Amro MM, Al-Mobarky A, Al-Homadhi ES (2007) Improved oil recovery by application of ultrasound waves to waterflooding. In: SPE 105370, presented at the 15th SPE Middle East Oil and Gas Show and Conference held in Bahrain International Exhibition Centre, Kingdom of Bahrain, 11–14 March
2. Beresnev IA, Vigil D, Li W (2005) Elastic waves push organic fluids from reservoir rock. Geophys Res Lett 32:L13303. doi:10.1029/2005gl023123CrossRef

3. Bjerknes VF (1906) Fields of Force. Columbia University Press, New York

4. Chen WI (1969) Influence of ultrasonic energy upon the rate of flow of liquids through porous media. Chemical Engineering. PhD thesis, West Virginia University, p 141

5. Duhon RD, Campbell JM (1965) SPE 1316—the effect of ultrasonic energy on flow through porous media. In: Second Annual Eastern Regional Meeting of SPE/AIME, Charleston, WV

6. Fairbanks HV, Chen WI (1971) Ultrasonic acceleration of liquid flow through porous media. Chem Eng Prog Symp Ser 67:108

7. Ganiev RF, Ukrainskii LE, Frolov KV (1989) Wave mechanism for the acceleration of a liquid flowing in capillaries and porous media. Sov Phys Doklady 34:267–283

8. Hamida T, Babadagli T (2007) Fluid-fluid interaction during miscible and immiscible displacement under ultrasonic waves. Eur Phys J B 60:447–462. doi:10.1140/epjb/e2008-00005-5CrossRef

9. Kostrov S, Wooden BO (2001) Mechanisms, flied suitability, and case studies for enhancement of oil recovery and production using in situ seismic stimulation. international conference elastic wave effect on fluid in the porous media. In: International symposium on nonlinear acoustics, Moscow

10. Simkin EM (1993) A possible mechanism of vibroseismic action on an oil-bearing bed. J Eng Phys Thermophys 64(4):355–359CrossRef

11. Simkin EM, Surguchev ML (1991) Advanced vibroseismic technique for water flooded reservoir stimulation, mechanism and field tests results. In: The 6th European IOR Symposium, Stavanger, Norway

CHAPTER 8

Mode & mechanism of low intensity pulsed ultrasound (LIPUS) in fracture repair

Andrew Harrison, Sheldon Lin, Neill Pounder, Yuko Mikuni-Takagaki

HIGHLIGHTS

- A mode and mechanism on how LIPUS enhances fracture repair is proposed.
- LIPUS drives the production of bone specific proteins to enhance bone healing.
- Integrins, on the surface of cells detect the pressure wave caused by LIPUS.
- The key gene which is upregulated after LIPUS exposure is COX2.
- COX2 is critical for LIPUS enhanced production of bone specific proteins.

ABSTRACT

It has been 30 years since the first level one clinical trial demonstrated low intensity pulsed ultrasound (LIPUS) could accelerate fracture repair. Since 1994 numerous investigations have been performed on the effect of LIPUS. The majority of these studies have used the same signal parameters comprised of an intensity of 30 mW/cm^2 SATA, an ultrasound carrier frequency of 1.5 MHz, pulsed at 1 kHz with an exposure time of 20 minutes per day. These studies show that a biological response is stimulated in the cell which produces bioactive molecules. The production of these molecules, linked with observations demonstrating the enhanced effects on mineralization by LIPUS, might be considered the general manner, or mode, of how LIPUS stimulates fractures to heal.

We propose a mechanism for how the LIPUS signal can enhance fracture repair by combining the findings of numerous studies. The LIPUS signal is transmitted through tissue to the bone, where cells translate this mechanical signal to a biochemical response via integrin mechano-receptors. The cells enhance the production of cyclo-oxygenese 2 (COX-2) which in turn stimulates molecules to enhance fracture repair. The aim of this review is to present the state of the art data related to LIPUS effects and mechanism.

KEYWORDS

Fracture; LIPUS; Ultrasound; Mechanism; Bone morphogenic protein; Cyclo-oxygenase 2

1. INTRODUCTION

Fracture healing is a complex process. Clinically, it requires adequate reduction of the displaced fracture and stabilization. The biological process consists of several distinct temporal phases. These phases are the inflammatory phase with cell proliferation, the chondrogenic phase with cartilage hypertrophy and angiogenesis, and the osteogenic phase with replacement of cartilage with woven bone and subsequent remodeling. During the fracture healing process, various cell types must interact and deliver adequate and appropriate levels of inflammatory and bioactive molecules. Despite the complexity, most fractures heal without a problem. In the United States, over eight million fractures occur annually, with a delayed or non-union rate of 5–10% [1].

Several risk factors are associated with impaired bone healing. These factors include the presence of systemic disease, such as diabetes [2]; advanced age [3] and [4]; or external chemical factors, such as alcohol abuse [5], smoking [6], or prescribed medications like chronic steroid use or chemotherapeutics [7]. These risk factors provide an ongoing clinical challenge for the treatment of fractures in our aging, comorbid population.

Several biophysical agents have been proposed and analyzed as therapeutic adjuncts to address this impaired bone healing process. These adjuncts include pulsed electromagnetic fields, combined magnetic fields, direct electrical current and low-intensity pulsed ultrasound. The pulsed electromagnetic systems generate asymmetric, quasirectangular electromagnetic pulses grouped in bursts. Individual pulse duration is approximately 260 μs, burst frequency is 15 Hz and the peak field strength is 20 mT. The nonunion effectiveness rate for patients using one of the pulsed electromagnetic system for an average of 7.1 hours per day was 80% [8]. Another pulsed electromagnetic system demonstrated a nonunion success rate of 63.5% for treatment periods up to 10 hours [9]. The external combined magnetic field device produces a static 20 μT magnetic field and a sinusoidal dynamic field at a frequency of 76.6 Hz and amplitude of ±40 μT. Patients use the device for 30 minutes per day, providing a nonunion heal rate of 60.7% [10]. The implanted direct current stimulator delivers a 20 μA current via a wire cathode continuously for 24 hours a day for at least 6 months, resulting in a long term nonunion success rate between 38.8% and 66.7% [11]. Low-intensity pulsed ultrasound (LIPUS) is indicated for certain fresh fractures and the treatment of nonunions. A 20 minute treatment time results in a 38% acceleration of fresh fractures [12] and [13] and an 86% nonunion success rate [14]. This review discusses LIPUS exclusively. When defining low intensity for LIPUS we will deal specifically with an intensity of 30 mW/cm^2 SATA (spatial average–temporal average) and a power of 117 mW delivered via an unfocussed ultrasound transducer with an active area of 3.88 cm^2, as measured by the radiation force balance method. The ultrasound carrier frequency is 1.5 MHz, pulsed at 1 kHz. The ultrasound power is measured by the Ohmic Instruments ultrasound power meter.

The use of low intensity ultrasound to accelerate the fracture repair process in humans was first reported by Xavier and Duarte in 1983 [15]. This success led to clinical trials in the United States [12] and [13] and the 1994 approval of EXOGEN by the U.S. Food and Drug Administration for the accelerated healing of certain fresh fractures. The product was approved for the treatment of established non-unions in 2000 [14] and [16]. All the LIPUS articles stated in this review have used the clinically approved ultrasound signal. Specifically, 1.5 MHz sinusoidal waves modulated in bursts of 200 ms at a repetition frequency of 1 kHz, and a spatial average–temporal average intensity of 30 mW/cm^2.

The positive effect of low-intensity pulsed ultrasound on the acceleration of bone healing has been demonstrated in non-clinical and clinical studies. Clinical studies have demonstrated significant positive effect of LIPUS in the treatment of fresh fractures[12] and [13] and non-unions [14], [16] and [17]. The prospective, randomized, multi-center, double-blind level I clinical studies investigating LIPUS treatment of conservatively treated tibia [12] and distal radius [13] both demonstrated 38% acceleration in fracture repair compared to fractures treated with a placebo device ($p < 0.0001$), as assessed radiographically and clinically by blinded physicians. For nonunion fractures, the studies [14], [16] and [17] included nonunion fractures that would not otherwise have healed. The fractures were at least 9 months old, with a minimum of 3 months since the last intervention. The only treatment change was daily LIPUS use. Gebauer et al. investigated the effect of LIPUS in 67 subjects, demonstrating heal rates of 85% after 5.6 months [16], while Nolte et al. studied 29 nonunions that resulted in an 86% heal rate in an average time of 5.5 months [14]. Neither study reported any safety issues.

The diffraction pattern of the ultrasound waves produced by a piston transducer, such as the LIPUS transducer, has two characteristic zones: the near field (close to the transducer) and the far field (farther from the transducer). The interface between the near field and far field is described by the equation:

$$Z = \frac{a^2 f}{c}$$

where Z is the near field length, a is the radius of the transducer after the near field length, f is the frequency, and c is the speed of sound in the medium [18]. The LIPUS transducer has a radius of 11 mm, emitting 1.5 MHz ultrasound wave and propagating in average tissue at 1540 ms^{-1}[19]. The computed value of Z is 118 mm. The ultrasound beam pattern measured in water using a hydrophone shows rapidly changing intensity amplitude across the beam in the near field ($Z = 0$ mm). The complexity (successive axial maxima and minima) of the ultrasound field decreases as the axial distance from the transducer increases. At $Z = 60$ mm (mid-near field), the complexity of the beam decreases and forms less maxima and minima. At $Z = 118$ mm (far field), the ultrasound pressure wave is more uniform. For the experiments described in this review, all the cell culture and in vivo experiments were performed in the near field. This is contrary to recommendations to place the in vitro cell culture within the far field where the pressure levels and temporal variation of intensity are controlled [20]. However, placement within the near field more closely matches the clinical situation as the majority of bones have little soft tissue coverage. The deepest bone treated clinically is the femur. With the far field starting at 118 mm, and assuming the femur is in the center of the thigh, the circumference of a patient's thigh would need to be 74 cm (29 inches) before a femur would be receiving ultrasound in the far field. Anthropometric data indicates that 95% of Americans have a mid-thigh circumference of less than 75 cm (29.5 inches) [21]. Therefore, application of the ultrasound coupled to a petri dish or directly to a rodent's limb provides a more clinically relevant ultrasound field. In the in vivo studies, the animals were sedated daily and the active or placebo transducer coupled to the skin with ultrasound gel directly above the fracture, unless otherwise stated. The in vivo models typically used either rats or mice, with a closed femoral fracture first described by Bonnarens and Einhorn [22]. This model is widely regarded as the standard in vivo model for traumatic fractures. The phases of fracture healing match those observed clinically, albeit on a shorter timescale. The consistency of the fracture allows this in vivo model to be used to investigate different interventions and mechanisms. The purpose of this review is to provide our current understanding regarding the mechanism and mode of action for LIPUS.

Routinely the literature states that 'the mechanism of LIPUS has yet to be elucidated'. We believe that there is evidence supporting both a credible mode and a mechanism behind the LIPUS technology. 'Mode' is described in dictionaries as 'a way in which something occurs or is done' [23]. Therefore the mode of action that we propose for how LIPUS stimulates an enhanced healing process to accelerate fracture repair can be seen as the stimulation of processes and molecules that contribute to heal the fracture is related to mechanoreceptors at the cellular level. The mechanism of how LIPUS achieves this is quite different from the mode. The dictionary definition of 'Mechanism' is 'the way in which something works or is brought about' [23]. This review will seek to investigate the evidence for both the mode and mechanism behind the LIPUS technology, first by explaining the osteogenic action of LIPUS on fracture repair, then by demonstrating the mechanism by evaluating the LIPUS signal and how this is 'interpreted' by the cells to elicit a healing response, using the best key evidence. This review is an update of a previous review in this journal in 2008 to pull together aspects of the biology to describe a mechanism of action [24]. Previously, LIPUS was believed to work equally in all phases of fracture repair. This update provides more detailed evidence supporting the impact on endochondral ossification and a clear effect on bone resorption during healing. In addition, there is a clear mechanistic link between integrin activation and COX2 upregulation.

2. THE MODE OF ACTION OF LIPUS IN FRACTURE REPAIR

2.1. Effect of LIPUS in animal models of fracture repair

In order to understand the effect LIPUS has on fracture repair, numerous pre-clinical models have been used to simulate the clinical situation. One such experimental in vivo model was performed treating fractures at different periods of repair. A rat closed femoral fracture was created in both hind limbs with the right femur exposed to LIPUS, and the left femur used as a control [25]. Rats were divided into four groups according to timing and duration of treatment. The animals in each group were treated with LIPUS for 20 minutes per day as follows: LIPUS treatments were performed in the Phase 1 group for 8 days, from day 1 to 8 after fracture; in the Phase 2 group for 8 days, from day 9 to 16 after fracture; in the Phase 3 group for 8 days, from day 17 to 24 after fracture. The last group was treated from days 1 to 24, throughout the healing process. The authors considered this to represent the hematoma, soft callus and mineralization phases of fracture repair. Animals were euthanized on day 25, and radiographs and torsional biomechanical testing were used to assess fracture healing. The maximal torque and stiffness in torsion of the fractured femur on the LIPUS-treated side was significantly higher than that of the contralateral control side in all groups ($p < 0.01$). This indicated that even partial treatment with LIPUS during Phase 1, 2, or 3 improved the mechanical properties of the fracture callus; however, treatment throughout the 24 days was most effective ($p < 0.05$). The histology presented in this study showed an acceleration of endochondral ossification, in that the LIPUS-treated fractures at day 25 resulted in much more bone formed than in controls at the same time point. This observation of accelerated endochondral ossification was reproduced in another in vivo study using a closed transverse mid-diaphyseal femoral fracture also in rats [26]. LIPUS was applied for 10 min daily, with the contralateral control limb receiving an inactive transducer. The formation of the new tissue after fracture was assessed by micro CT each week for three weeks, with phasing thresholds which differentiated un-mineralized tissue, less dense bone, calcified cartilage, and dense or cortical bone. This technique provided further evidence of LIPUS influencing the fracture cascade, including enhanced bone formation in the fracture gap and enhanced resorption of the 'old' cortical bone. Using histology, this study also showed that the number of osteoclasts was significantly increased in the LIPUS-treated fractures compared with controls, which suggests that LIPUS can influence the remodeling phase of fracture repair. The stimulation of osteoclastic action has been hypothesized as resulting from increased concentrations of ATP released by LIPUS-treated osteoblasts into culture medium which was associated with increased Receptor Activator of Nuclear Factor Kappa B Ligand (RANKL) and decreased osteoprotegerin expression [27].

These data suggest, along with other studies [25], [26] and [28], that the most significant effect of LIPUS on fracture repair is on endochondral ossification, the process of converting soft, cartilaginous callus to hard, mineralized callus, increasing the mechanical stability of the healing fracture. For an efficient process of endochondral ossification, it is accepted that the fracture site needs to be vascularized similarly to that observed in growing bone [29].

2.2. Processes in accelerating endochondral remodeling by LIPUS

As seen in the preclinical models of fracture repair, LIPUS enhances the process of endochondral ossification. One of the key processes in endochondral remodeling is the increase in blood vessels invading the fracture site. LIPUS has been shown to increase key growth factors involved in the process of angiogenesis, the formation of new blood vessels. Vascular endothelial growth factor (VEGF) is a growth factor that causes the formation of new blood vessels [30]. An in vivo study investigated the effect of LIPUS in a type 1 diabetes mellitus (DM) femoral fracture rat model [31]. The number of blood vessels invading the fracture site was assessed by measuring PECAM 1 levels. In the DM group the density of blood vessels at the fracture site was significantly reduced compared with normal control animals ($p = 0.004$). With the addition of LIPUS, the DM animals had a significantly increased blood vessel density compared with DM animals that had not received LIPUS ($p = 0.017$), representing a 77.6% increase. This effect was linked to significant increases in VEGF production in LIPUS-treated fractures ($p = 0.02$).

The impact of LIPUS on the process of endochondral ossification was clearly demonstrated in a mid-diaphysis femur fracture model in aged mice [32]. At day 21 post-fracture in 8 week old mice, the process of endochondral ossification was mostly complete. However, in aged, 40 week old mice, the fracture gap was still present at 21 days post-fracture and the tissue that remained was cartilaginous. The impact of age on the fracture repair process was diminished when the animals were treated with LIPUS. In the 40 week old mice, radiographs showed that the fracture callus after 21 days of LIPUS treatment was fully mineralized, leading to greater stability of the fracture. The onset of endochondral ossification occurred at the same time in the aged animals, but the overall process was accelerated. In a parallel study of aged (1 year old) mice with bilateral femur factures, X-rays and histology showed a similar phenomenon. The mice were treated for 20 minutes per day with an active or placebo ultrasound transducer. The period of endochondral ossification in control aged mice was considerably longer compared to the young mice, approximately 18 days compared to 7 days over the 28 day assessment period, with LIPUS shortening the endochondral ossification in aged mice to approximately 11 days [33].

2.3. Osteogenic response of bone cells

In the process of endochondral ossification, the soft cartilaginous or fibrous callus is converted to a calcified hard callus. In order to complete this process, the cells in the local environment produce factors that are associated with this conversion, such as type X collagen, alkaline phosphatase and osteocalcin [34].

Two papers have demonstrated the enhanced stimulation of osteogenic cells by LIPUS to drive endochondral ossification [35] and [36]. These two independent research teams both took human osteogenic cells and exposed them to LIPUS in vitro. In both sets of experiments LIPUS was applied through the bottom of the culture plates for 20 minutes daily at 37 °C and therefore the cells experienced a near field exposure. The same ultrasound transducer that is used clinically, described previously, was used in both sets of experiments. Periosteal cells [35] and isolated cells from fractures in the early phase of fracture repair [36] were used. The results from these two studies were remarkably similar, with alkaline phosphatase levels stimulated by LIPUS increasing with time in culture over controls. In the case of the human periosteal cells [35], 20 minutes exposure of LIPUS over 4 days resulted in a significant difference in alkaline phosphatase over controls of around 10% ($p = 0.01$) and 30% ($p = 0.03$) on days 2 and 4 respectively. With progenitor cells isolated from the fracture hematoma [36], LIPUS was applied daily for 20 minutes over 28 days. Alkaline phosphatase levels were significantly higher after 2, 4, 7, and 14 days of LIPUS exposure ($p < 0.05$) over controls, showing a 20% increase at day 2 through to, 40% on day 14; the same was true with regard to osteocalcin secretion. Two master control genes of the osteogenic lineage, specifically cbfa-1/RUNX2 and osterix, were significantly upregulated compared to controls ($p < 0.05$). The increase in alkaline phosphatase levels, together with osteocalcin secretion and expression of other osteoblast related genes demonstrated that LIPUS promoted osteogenic differentiation of the progenitor cells. The net result of this activity was to increase calcium deposition in the cultures of cells as seen by alizarin red staining. These two studies show strikingly comparable results demonstrating osteogenic differentiation in human cells in response to LIPUS. These results also correspond to what is seen in animal models and other pre-clinical studies, indicating that the responses initiated in the human is similar to that seen in the animal. Increases in bone specific genes was also observed in vivo following the extraction of mRNA from the fracture callus of aged mice at 7 days post fracture [33]. LIPUS treatment led to increases in bone morphogenic proteins BMP-2, BMP-4 and BMP-6 of around 8, 10 and 5-fold higher respectively over controls. This enhanced expression of BMPs in response to LIPUS has also been reported in a study of ROS 17/2.8 osteoblastic cells stimulated in 6 well tissue culture plates [37]. The ultrasound stimulation configuration differed from previous in vitro studies, in that the cells were stimulated from above with the transducer touching the surface of the medium, with the cells 3–4 mm away from the surface of the transducer. Although the cells were stimulated from above, they were still within the near ultrasound field. The cells were stimulated with ultrasound 20 minutes per day for up to 7 days. After 7 days stimulation in culture there were clear increases in protein levels for BMPs 2, 4 and 7 in the LIPUS treated cells versus control cultures, as determined by Western blot analysis.

Another role of enhanced vascularity stimulated by the production of angiogenic growth factors is in recruiting osteogenic progenitor cells to the fracture site from the systemic circulation. To evaluate this effect, parabiotic mice were produced by surgically joining the circulatory systems of two mice, one of which constitutively expressed green fluorescent protein (GFP) in all cells and tissues and a syngeneic wild-type mouse [38]. Femoral fractures were then created in the wild type (non-GFP) animals and treated with LIPUS or a sham device for 20 minutes per day for up to 4 weeks. Any GFP-positive cells in the callus of a wild-type animal would, by definition, be derived from circulating progenitor cells. At two and four weeks, animals were euthanized and the number and localization of GFP-positive cells were assessed. The hard callus area was significantly greater in the LIPUS group at 2 and 4 weeks post fracture ($p < 0.05$), as reported in other studies [25], [26] and [28]. In addition, a significant increase in alkaline phosphatase positive cells in the fractures treated with LIPUS at 2 weeks post fracture was observed ($p < 0.05$), which complements the increased mRNA expression of alkaline phosphatase reported in other studies [39]. Fractures treated with LIPUS had significantly more GFP cells, from the adjacent animals as compared to control fractures, as assessed by histomorphometry, suggesting that LIPUS increased the homing of the progenitor cells to the fracture site. The authors investigated the mechanism for the enhanced recruitment of osteogenic progenitors. Confocal micrographs of the fracture site at 2 weeks showed many CXCR4 positive cells in the LIPUS treated fractures that were virtually absent in control fractures. CXCR4 is expressed on the surface of cells and is the receptor for stromal derived factor 1 (SDF-1 or CXCL12) which plays a key role in the recruiting of circulating cells to sites of repair [40] and [41]. In a separate study [42] the migration of transplanted mesenchymal stem cells (MSCs) to the fracture site was investigated by fluorescent imaging. MSCs were isolated from a rat femur and stimulated with LIPUS for 20 minutes per day for 3 days. The cells were subsequently injected into rats with a closed femoral fracture. This study demonstrated that LIPUS promoted MSC migration to the fracture site, which was associated with an increase of local and serum SDF-1 level. The in vitro analysis showed that LIPUS upregulated SDF-1 and CXCR4 expressions in MSCs. MSCs migration was promoted by LIPUS. Therefore, two studies independently concluded that LIPUS is having an effect of increasing the recruitment of cells to the fracture site [38] and [42].

3. MECHANISM OF ACTION OF LIPUS IN FRACTURE REPAIR

3.1. Process of LIPUS stimulation

As described earlier the general manner of how LIPUS enhances fracture repair can be explained by the mode of action. The mode is the expression of molecules and actions that need to occur in order to achieve fracture repair, as outlined above. However, to truly understand the process of how LIPUS influences fracture repair we must understand the mechanism of action, which is the description of the events after the signal leaves the transducer, enters the tissue and initiates key events that lead to fracture repair.

Work performed on human cadaveric specimens has demonstrated that the LIPUS signal can produce motion at the fracture site [43]. An osteotomy was performed on the radius of the cadaver, LIPUS applied, and the motion measured using a laser interferometer. Motion was measured at the proximal and distal edges of the bone in the osteotomy, while the LIPUS device was activated (Fig. 1). From these experiments it was found that motion occurred at a frequency of 1 kHz, matching the pulse modulation of the LIPUS signal. The interferometer measured velocity amplitude. Velocity amplitude measurements can be converted to displacement amplitudes using the following equation:

$$v = A\cos \omega t = A\cos[2\pi ft]$$

$$s = \int v \cdot dt = \int A\sin[2\pi ft] = \frac{A}{2\pi f}\sin[2\pi ft]$$

where A = velocity amplitude, s = displacement amplitude, f = measurement frequency. When calculated, it was found that at a measurement frequency of 1 kHz, the velocity amplitude of the bone ends and soft tissue ranged from 1 μm s^{-1} to 3.5 μm s^{-1}, equating to 0.16–0.56 nm displacement, respectively. It is understood that bone responds to mechanical stimulations and the displacement is referred to as micromotion. The normal levels of displacement associated with micro-motion are accepted to be in the range of 0.5–2 mm. Therefore the displacements caused by LIPUS are around 1000 times less than these values, which would suggest that LIPUS has a different effect on bone compared to micro-motion. Micro-motion has an anabolic effect on bone mass [44], whereas LIPUS does not. This difference was demonstrated clinically when patients suffering from osteoporosis had their distal radius treated daily with LIPUS for 3 months, and evaluated 3 months after discontinuing the treatment[45]. The results showed that the rate of bone change (trabecular bone mineral density and integral bone mineral density) did not significantly differ between the site treated with LIPUS and the contralateral control at either 3 months or 6 months ($p = 0.954$ and$p = 0.410$ respectively), suggesting that micro-motion and nano-motion have different effects.

The EXOGEN device, which produces low intensity pulsed ultrasound, was approved by the US Food and Drug Administration in 1994 to accelerate the healing of certain fresh fractures based on the intensity of 30 mW/cm^2 SATA. This intensity of 30 mW/cm^2 has been used for all the clinical studies and the majority of pre-clinical studies when LIPUS has been used. Studies have been performed that have reduced the intensity to see if a threshold could be detected, below which no biological effect occurred. Rat bone marrow stromal cells (rBMSC) were recovered from the femurs of adult animals and cultured in 12 well plates. The culture plates were placed on an ultrasound transducer array, with the LIPUS coming from below. The in vitro cell cultures were treated with ultrasound intensities of 2, 15, and 30 mW/cm^2. Alkaline phosphatase activity was measured in cells that were treated for 20 minutes per day for 3, 5, and 7 days and mineral incorporation over a 24 day period. Interestingly, all intensities, even as low as 2 mW/cm^2, showed increases in alkaline phosphatase activity and mineralization over the sham control [46]. These data indicate that it is important for the cells to receive the LIPUS signal, rather than a specific intensity. The analogy would be similar to a person speaking in a conversation; you can either whisper the message or shout the message, but as long as the recipient receives the message the volume is irrelevant. This in vitro conclusion is supported by an in vivo fracture repair study [47]. Using the closed rat femoral fracture model, the effect of fracture healing of LIPUS treatment at 30 mW/cm^2 (approved clinically) and 150 mW/cm^2 SATA was evaluated by weekly radiographs, micro CT, and biomechanically. The study concluded that LIPUS at I_{SATA} of 150 mW/cm^2 did not further enhance fracture healing over that of 30 mW/cm^2 and was closer to that of control values compared to that of 30 mW/cm^2.

3.2. Conversion of mechanical signal into biochemical signaling

For the ultrasound to have a biological effect, the mechanical wave must be converted to a biochemical signal within the cell. The mechano-receptors, integrins, have been implicated in this process. Independent groups have reported activation of integrin associated signaling pathways through the observation of focal adhesions forming on the surface of cells treated with LIPUS [48] and [49]. Focal adhesions are conglomerations of integrin molecules which initiate intra-cellular proteins to translocate to inside of the cell where the intra-cellular components of integrins are found. There are many components of the intracellular portion of the focal adhesion that are able to initiate a cell signaling cascade. One of the key focal adhesion proteins implicated with the transduction of the LIPUS signal from a mechanical force to a biochemical signal is focal adhesion kinase (FAK), which is phosphorylated in response to LIPUS. Several groups have also shown a link between the stimulation of integrins and extracellular signal-regulated kinase (ERK) signaling in the transmission of LIPUS, which is considered downstream of FAK[49] and [50].

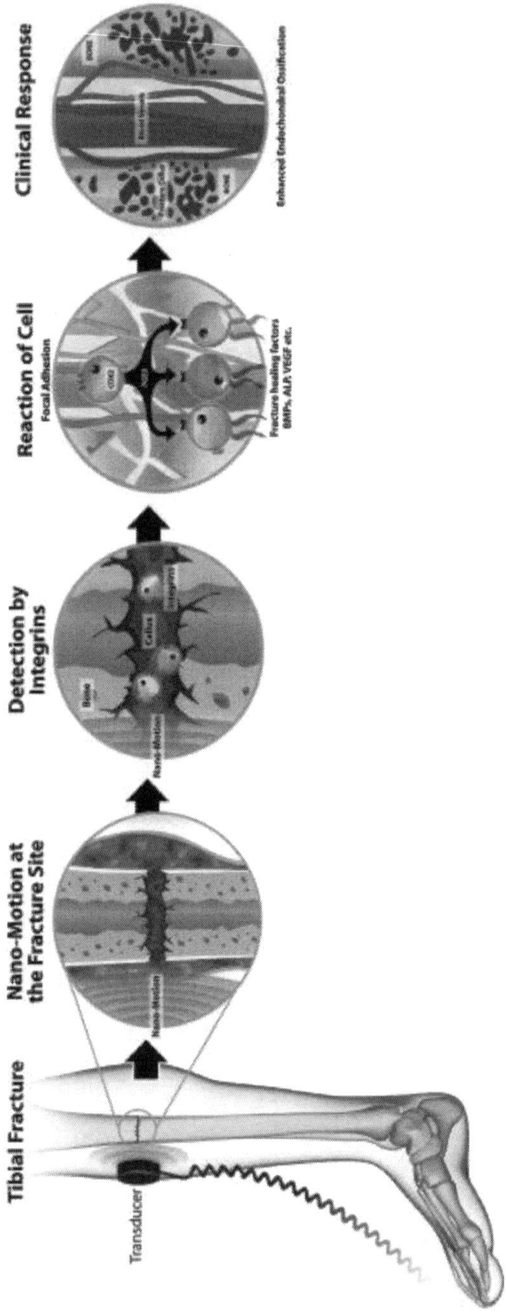

Figure. 1. The transducer is placed on the skin above the fracture and the low intensity pulsed ultrasound produces nano-motion at the fracture site. The nano-motion is detected by integrins and the biomechanical wave is converted into a biochemical wave in the cell.

One of the main actions in the cell is the production of cyclo-oxygenase 2 or COX2. The enzymatic action of COX2 is the rate limiting step in the production of prostaglandin E2 or PGE2, which is released from the cells and interacts with surrounding cells through their EP receptors. This action enhances the process of endochondral ossification, which is key for the clinical response.

3.3. Downstream effect of biochemical signaling: COX2 in the mechanism of LIPUS induced fracture repair

From the best available evidence, it is strongly suggested that a key molecule stimulated by LIPUS through the pathways described above is cyclo-oxygenase 2 (COX2) [51]. COX2 is the rate-limiting enzyme in the production of prostaglandin E2 (PGE2). It has also been demonstrated that LIPUS stimulates prostaglandin H synthase 2 (PGHS-2) expression and PGE2 synthesis [52]. The inhibition of COX2 by non-steroidal anti-inflammatory drugs (NSAIDs) has been strongly implicated in impaired fracture repair in both clinical and pre-clinical studies [53] and [54]. It is also reported that older mice produce less COX2, leading to reduced capacity of fracture repair [4]. It is therefore unsurprising that a number of publications have shown that COX2 is a key molecule stimulated in cells after the application of LIPUS [50] and [55]. The secretion of PGE2 was significantly upregulated over 24 h after a single 20 min application of LIPUS[50] and [55]. One study [50] has linked all aspects of the signaling described above and associated this to the enhanced mineralization caused by LIPUS. This work demonstrated that LIPUS was able to increase the levels of integrin expressed on the cell surface (α2, α5, β1 and β5 sub-units). They also showed that LIPUS caused the phosphorylation of FAK, which led to the phosphorylation of the p85 sub-unit of phospho-inositol 3 kinase (PI3K), then the phosphorylation of AKT, and finally the translocation of the p65 subunit of NF-κβ to the nucleus. At each stage of this pathway, when inhibitors were used, the enhanced COX2 production was ablated, strongly suggesting that this intra-cellular pathway is stimulated by LIPUS to initiate the production of COX2. These results were corroborated using a COX2 luciferase assay which, after the treatment with LIPUS, significantly increased this activity compared to non-treated controls. When these LIPUS stimulated cells were treated with either chemical pathway inhibitors or dominant negative mutant forms of FAK, PI3K, AKT or ERK, the activity of the COX2 was significantly reduced. Finally, it was demonstrated that, via this pathway, LIPUS could enhance the mineralization of osteoblastic cells. This enhanced mineralization stimulated by LIPUS could be significantly ablated when the cultures were treated with dominant-negative mutant forms of FAK, PI3K or AKT. However, the most compelling evidence that the stimulation by LIPUS through this pathway activates COX2 was shown when these cultures were treated with NS-398, a COX2 inhibitor. Again the enhanced mineralization caused by LIPUS was significantly reduced. This evidence strongly implicates the LIPUS signal is transduced through these specific pathways in the cell, which leads to the production of COX2 and drives mineralization in stimulated osteoblasts.

In vitro cell biology experiments, especially in cells of established cell lines, can only explain so much, and the mechanism of COX2 activation was verified in an in vivo model. In femur fractures of standard aged (one year old) wild type mice, stimulation with LIPUS showed an enhanced healing response compared to untreated contralateral controls, as previously discussed. However, in animals where COX2 had been knocked out, fracture healing via histology was undistinguishable compared to controls. These results demonstrate that the production of COX2 is essential for the induction of gene transcription for both osteogenesis and remodeling, and thereby for the enhanced fracture repair stimulated by LIPUS [33]. In a study of alveolar bone defect as a result of molar tooth removal in rats, involvement of humoral PGE2, which was released from the site of LIPUS stimulation via upregulated COX2, was suggested in the de novo expression of CXCR4 in the remote bone marrow in tibia [56].

COX2 is associated with the initiation of inflammatory responses by producing prostaglandins. PGE2 stimulates osteoclast formation by increasing RANKL mRNA expression via cAMP-protein kinase A (PKA) pathways in osteoblasts. Therefore, a timely COX2 induction by LIPUS inevitably accelerates callus remodeling from cartilage and woven bone into lamellar bone to restore the cortical bone. The activation of resorbing cells (osteoclast and chondroclast) is supported by the increased RANKL expression, both at the protein and mRNA levels in established cell lines and osteoblastic cells [57] and [58]. Inflammatory conditions may activate multiple pathways in inflammasome preparation as a first step priming through toll-like receptors and TNF or IL-1β receptors that also activate NF-kB [59].

4. SUMMARY

LIPUS waves cause nanomotion at the fracture site. The mechanical signal is converted to a biochemical signal inside the cells and is ultimately transmitted through signaling molecules to drive the production of COX2 in the cell. This cascade requires the involvement of integrins and the formation of focal adhesions. The enhanced COX2 stimulation, likely through the production of PGE2, drives the expression of osteogenic genes. These osteogenic genes stimulate enhanced mineralization, which is seen as enhanced endochondral ossification that heals the fracture (Fig. 2). This enhanced mineralization caused by LIPUS can be observed in cell culture experiments [35],[36] and [50], in vivo models of fracture repair [25], [26] and [33] and in clinical trial data[60].

5. CONCLUSION

Many intensities and frequencies have been evaluated, but the greatest number of clinical and pre-clinical studies have used a signal comprised of 30 mW/cm^2 I_{SATA}, an ultrasound frequency of 1.5 MHz and pulsed at 1 kHz. It is this signal that was used in the level-I clinical studies demonstrating acceleration of fracture repair in diaphyseal tibial fractures and distal radial fractures respectively [12] and [13]. Studies have also demonstrated that this signal is highly efficacious in the treatment of recalcitrant non-union fractures [14] and [16].

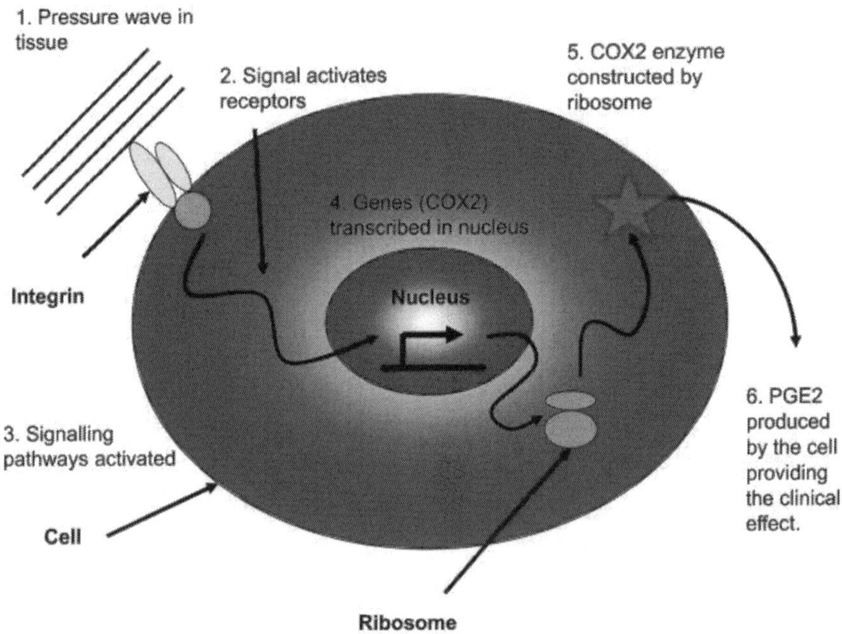

Figure. 2. As the pressure wave comes through the tissues it interacts with cells and is detected by integrins. Multiple pathways are initiated in the cell which leads to the transcription of genes in the nucleus, the main activation is in the transcription of COX2. The mRNA of COX2 is converted to protein that then produces the eicosanoid prostaglandin E2 which is released from the cells to have the clinical effects.

A search of the PubMed database of LIPUS finds over 250 articles. However, most articles generally state that the underlying mechanism of action remains unclear. From the data presented here, we hope to convince the reader that there is a rich literature in the science of LIPUS and that these articles indicate that there is a robust mode and mechanism behind the LIPUS technology.

It is commonly understood that fracture repair can be described as different phases. These phases include inflammatory, intramembranous ossification, chondrogenesis, endochondral ossification and finally remodeling. It is not clear at this time exactly which types of cells are involved in each phase; however it has been demonstrated that LIPUS can have positive effects in all these phases [25].

The central molecule in this proposed mechanism of action is cyclo-oxygenase 2 (COX2), the key enzyme in the production of prostaglandins. Prostaglandins have been shown to play an important role in bone fracture repair [53] and [54]. This evidence could indicate, therefore, that stimulating COX2 might be a good attribute for fracture repair. Studies have shown a positive effect on COX2 and PGE2 levels in response to LIPUS[50] and [55]. However, the most compelling evidence of this was the near total lack of efficacy of LIPUS in the aged COX2 knockout mouse, which strongly implies that COX2 sits centrally in the mechanism of action of LIPUS [33].

In conclusion, we have set out in this paper to demonstrate that within the scientific literature there is evidence that suggests a mechanism of action of how LIPUS stimulates fracture repair. We have illustrated how the signal is transmitted through tissue to the fracture site, how the bone at the fracture site is made to move at the nano-meter range, and how cells are able to sense this signal by forming focal adhesions. We have shown how signaling pathways are initiated within the cell and how COX2 is produced leading to increased production of PGE2, which when blocked correlates to a lack of effect of LIPUS in enhanced fracture repair. The LIPUS signal has had a long history of safe and efficacious use and with continuing investigation more information will be generated in how this technology can stimulate the healing of tissues.

ACKNOWLEDGEMENTS

Andrew Harrison and Neill Pounder are employees of Bioventus that manufactures and sells a LIPUS device known as EXOGEN.

REFERENCES

1. A. Praemer, S. Furner, D.P. Rice, Musculoskeletal conditions in the United States, Am. Acad. Orthop. Surg. (1999) 85–124.

2. L. Oei, F. Rivadeneira, M.C. Zillikens, E.H. Oei, Diabetes, diabetic complications, and fracture risk, Curr. Osteoporos. Rep. 13 (2) (2015) 106–115.

3. J.D. Heckman, J. Sarasohn-Kahn, The economics of treating tibia fractures. The cost of delayed unions, Bull. Hosp. Joint Dis. 56 (1) (1997) 63–72.

4. A.A. Naik, C. Xie, M.J. Zuscik, P. Kingsley, E.M. Schwartz, H. Awad, R. Guldberg, H. Drissi, J.E. Puzas, B. Boyce, X. Zhang, R.J. O'Keefe, Reduced COX-2 expression in aged mice is associated with impaired fracture healing, J. Bone Miner. Res. 24 (2) (2009) 251–264.

5. M.S. Gaston, A.H. Simpson, Inhibition of fracture healing, J. Bone Joint Surg. Br. 89 (12) (2007) 1553–1560.

6. J.A. Bishop, A.A. Palanca, M.J. Bellino, D.W. Lowenberg, Assessment of compromised fracture healing, J. Am. Acad. Orthop. Surg. 20 (5) (2012) 273– 282.

7. E.M. Kagel, T.A. Einhorn, Alterations of fracture healing in the diabetic condition, Iowa Orthop. J. 16 (1996) 147–152.

8. D.E. Garland, B. Moses, W. Salyer, Long-term follow-up of fracture nonunions treated with PEMFs, Contemp. Orthoped. 22 (3) (1991) 295–302.

9. The Biomet□ EBI Bone Healing System, Physician Manual and Package Insert, 2013, (accessed 17.02.2016).

10. DJO Global CMF OL1000TM Product Information, (accessed 17.02.2016).

11. EBI OsteoGen Surgically Implanted Bone Growth Stimulator Patient Risk Information, 2004, (accessed 17.02.2016).

12. J.D. Heckman, J.P. Ryaby, J. McCabe, J.J. Frey, R.F. Kilcoyne, Acceleration of tibial fracture-healing by non-invasive, low-intensity pulsed ultrasound, J. Bone Joint Surg. Am. 76A (1) (1994) 26–34.

13. T.K. Kristiansen, J.P. Ryaby, J. McCabe, J.J. Frey, L.R. Roe, Accelerated healing of distal radial fractures with the use of specific, low-intensity ultrasound, J. Bone Joint Surg. Am. 79A (7) (1997) 961–973.

14. P.A. Nolte, A. van der Krans, P. Patka, I.M.C. Janssen, J.P. Ryaby, G.H.R. Albers, Low-intensity pulsed ultrasound in the treatment of nonunions, J. Trauma 51 (4) (2001) 693–703.

15. C.A.M. Xavier, L.R. Duarte, Ultrasonic stimulation of bone callus: clinical applications, Rev. Brasileira Orthop. 18 (3) (1983) 73–80.

16. D. Gebauer, E. Mayr, E. Orthner, J.P. Ryaby, Low-intensity pulsed ultrasound: effects on nonunions, Ultrasound Med. Biol. 31 (10) (2005) 1391–1402.

17. X. Roussignol, C. Currey, F. Duparc, F. Dujardin, Indications and results for the ExogenTM ultrasound system in the management of non-union: a 59-case pilot study, Orthop. Traumatol. Surg. Res. 98 (2) (2012) 206–213.

18. M.C. Ziskin, Specification of acoustic output level and measurement uncertainty in ultrasound exposimetry, IEEE Trans. Ultrason. Ferroelectr. Freq. Control 50 (2003) 1023–1034.

19. F.A. Duck, Physical Properties of Tissue: A Comprehensive Reference Book, Academic Press, London, 1990.

20. F. Padilla, R. Puts, L. Vico, K. Raum, Stimulation of bone repair with ultrasound: a review of the possible mechanic effects, Ultrasonics 54 (2014) 1125–1145.

21. M.A. McDowell, D.D. Dryer, C.L. Ogden, K.M. Flegal, Anthropometric Reference Data for Children and Adults: United States, 2003–2006, National Health Statistics Reports No. 10, National Center for Health Statistics, Hyattsville, MD, USA, 2008.

22. F. Bonnarens, T.A. Einhorn, Production of a standard closed fracture in laboratory animal bone, J. Orthop. Res. 2 (1) (1984) 97–101.

23. The Oxford English Dictionary, third ed., Oxford University Press, Oxford, 2008.

24. N.M. Pounder, A.J. Harrison, Low intensity pulsed ultrasound for fracture healing: a review of the clinical evidence and the associated biological mechanism of action, Ultrasonics 48 (4) (2008) 330–338.

25. Y. Azuma, M. Ito, Y. Harada, H. Takagi, T. Ohta, S. Jingushi, Low-intensity pulsed ultrasound accelerates rat femoral fracture healing by acting on the various cellular reactions in the fracture callus, J. BoneMiner. Res. 16 (4) (2001) 671–680.

26. T.A. Freeman, P. Patel, J. Parvizi, V. Antoci Jr., I.M. Shapiro, Micro-CT analysis with multiple thresholds allows detection of bone formation and resorption during ultrasound-treated fracture healing, J. Orthop. Res. 27 (5) (2009) 673– 679.

27. M.J. Hayton, J.P. Dillon, D. Glynn, J.M. Curran, J.A. Gallagher, K.A. Buckley, Involvement of adenosine 50 -triphosphate in ultrasound-induced fracture repair, Ultrasound Med. Biol. 31 (8) (2005) 1131–1138.

28. S.J. Wang, D.G. Lewallen, M.E. Bolander, E.Y.S. Chao, D.M. Ilstrup, J.F. Greenleaf, Low intensity ultrasound treatment increases strength in a rat femoral fracture model, J. Orthop. Res. 12 (1) (1994) 40–47.

29. T.K. Daftari, T.E. Whitesides Jr., J.G. Heller, A.C. Goodrich, B.E. McCarey, W.C. Hutton, Nicotine on the revascularization of bone graft, Spine 19 (8) (1994) 904–911.

30. M. Ehrbar, V.G. Djonov, C. Schnell, S.A. Tschanz, G. Martiny-Baron, U. Schenk, J. Wood, P.H. Burri, J.A. Hubbell, A.H. Zisch, Cell-demanded liberation of VEGF121 from fibrin implants induces local and controlled blood vessel growth, Circ. Res. 94 (8) (2004) 1124–1132.

31. M. Coords, E. Breitbart, D. Paglia, N. Kappy, A. Gandhi, J. Cottrell, N. Cedeno, N. Pounder, J.P. O'Connor, S.S. Lin, The effects of low-intensity pulsed ultrasound upon diabetic fracture healing, J. Orthop. Res. 29 (2) (2011) 181–188.

32. M. Katano, K. Naruse, K. Uchida, Y. Mikuni-Takagaki, M. Takaso, M. Itoman, K. Urabe, Low intensity pulsed ultrasound accelerates delayed healing process by reducing the time required for the completion of endochondral ossification in the aged mouse femur fracture model, Exp. Anim. 60 (4) (2011) 385–395.

33. K. Naruse, H. Sekiya, Y. Harada, S. Iwabuchi, Y. Kozai, R. Kawamata, I. Kashima, K. Uchida, K. Urabe, K. Seto, M. Itoman,

Y. Mikuni-Takagaki, Prolonged endochondral bone healing in senescence is shortened by low-intensity pulsed ultrasound in a manner dependent on COX-2, Ultrasound Med. Biol. 36 (7) (2010) 1098–1108.

34. G.S. Stein, J.B. Lian, L.G. Gerstenfeld, V. Shalhoub, M. Aronow, T. Owen, E. Markose, The onset and progression of osteoblast differentiation is functionally related to cellular proliferation, Connect. Tissue Res. 20 (1–4) (1989) 3–13.

35. K.S. Leung, W.H. Cheung, C. Zhang, K.M. Lee, H.K. Lo, Low intensity pulsed ultrasound stimulates osteogenic activity of human periosteal cells, Clin. Orthop. Relat. Res. 418 (2004) 253–259.

36. T. Hasegawa, M. Miwa, Y. Sakai, T. Niikura, M. Kurosaka, T. Komori, Osteogenic activity of human fracture haematoma-derived progenitor cells is stimulated by low-intensity pulsed ultrasound in vitro, J. Bone Joint Surg. Br. 91 (2) (2009) 264–270.

37. A. Suzuki, T. Takayama, N. Suzuki, T. Kojima, N. Ota, S. Asano, K. Ito, Daily lowintensity pulsed ultrasound stimulates production of bone morphogenetic protein in ROS 17/2.8 cells, J. Oral Sci. 51 (1) (2009) 29–36.

38. K. Kumagai, R. Takeuchi, H. Ishikawa, Y. Yamaguchi, T. Fujisawa, T. Kuniya, S. Takagawa, G.F. Muschler, T. Saito, Low-intensity pulsed ultrasound accelerates fracture healing by stimulation of recruitment of both local and circulating osteogenic progenitors, J. Orthop. Res. 30 (9) (2012) 1516–1521.

39. J. Unsworth, S. Kaneez, S. Harris, J. Ridgway, S. Fenwick, D. Chenery, A. Harrison, Pulsed low intensity ultrasound enhances mineralisation in preosteoblast cells, Ultrasound Med. Biol. 33 (9) (2007) 1468–1474.

40. M.S. Penn, SDF-1:CXCR4 axis is fundamental for tissue preservation and repair, Am. J. Pathol. 177 (5) (2010) 2166–2168.

41. S. Otsuru, K. Tamai, T. Yamazaki, H. Yoshikawa, Y. Kaneda, Circulating bone marrow-derived osteoblast progenitor cells are recruited to the bone-forming site by the CXCR4/stromal cell-derived factor-1 pathway, Stem Cells 26 (1) (2008) 223–234.

42. F.Y. Wei, K.S. Leung, G. Li, J. Qin, S.K. Chow, S. Huang, M.H. Sun, L. Qin, W.H. Cheung, Low intensity pulsed ultrasound enhanced mesenchymal stem cell recruitment through stromal derived factor-1 signaling in fracture healing, PLoS ONE 9 (9) (2014) e106722.

43. J.F. Greenleaf, R.R. Kinnick, J.T. Bronk, M.E. Bolander, Ultrasound induced tissue motion during fracture treatment?, Ultrasound Med Biol. 29 (5) (2003) S157– S158.

44. J. Wolff, The Law of Bone Remodelling (Das Gesetz der Transformation der Knochen, Hirschwald), Springer, Berlin, 1892.

45. K.S. Leung, W.S. Lee, W.H. Cheung, L. Qin, Lack of efficacy of low-intensity pulsed ultrasound on prevention of postmenopausal bone loss evaluated at the distal radius in older Chinese women, Clin. Orthop. Relat. Res. 427 (2004) 234– 240.

46. S.R. Angle, K. Sena, D.R. Sumner, A.S. Virdi, Osteogenic differentiation of rat bone marrow stromal cells by various intensities of low-intensity pulsed ultrasound, Ultrasonics 51 (3) (2011) 281–288.

47. C.H. Fung, W.H. Cheung, N.M. Pounder, F.J. de Ana, A. Harrison, K.S. Leung, Effects of different therapeutic ultrasound intensities on fracture healing in rats, Ultrasound Med. Biol. 38 (5) (2012) 745–752.

48. C.M. Mahoney, M.R. Morgan, A. Harrison, M.J. Humphries, M.D. Bass, Therapeutic ultrasound bypasses canonical syndecan-4 signaling to activate rac1, J. Biol. Chem. 284 (13) (2009) 8898–8909.

49. S. Zhou, A. Schmelz, T. Seufferlein, Y. Li, J. Zhao, M.G. Bachem, Molecular mechanisms of low intensity pulsed ultrasound in human skin fibroblasts, J. Biol. Chem. 279 (52) (2004) 54463–54469.

50. C.H. Tang, R.S. Yang, T.H. Huang, D.Y. Lu, W.J. Chuang, T.F. Huang, W.M. Fu, Ultrasound stimulates cyclooxygenase-2 expression and increases bone formation through integrin, focal adhesion kinase, phosphatidylinositol 3- kinase, and Akt pathway in osteoblasts, Mol. Pharmacol. 69 (6) (2006) 2047– 2057.

51. P. Reher, M. Harris, M. Whiteman, H.K. Hai, S. Meghji, Ultrasound stimulates nitric oxide and prostaglandin E2 production by human osteoblasts, Bone 31 (1) (2002) 236–241.

52. V. Saini, S. Yadav, S. McCormick, Low-intensity pulsed ultrasound modulates shear stress induced PGHS-2 expression and PGE2 synthesis in MLO-Y4 osteocyte-like cells, Ann. Biomed. Eng. 39 (1) (2011) 378–393.

53. P.V. Giannoudis, D.A. MacDonald, S.J. Matthews, R.M. Smith, A.J. Furlong, P. De Boer, Nonunion of the femoral diaphysis. The influence of reaming and nonsteroidal anti-inflammatory drugs, J. Bone Joint Surg. Br. 82 (5) (2000) 655– 658.

54. A.M. Simon, M.B. Manigrasso, J.P. O'Connor, Cyclo-oxygenase 2 function is essential for bone fracture healing, J. Bone Miner. Res. 17 (6) (2002) 963–976.

55. K. Naruse, A. Miyauchi, M. Itoman, Y. Mikuni-Takagaki, Distinct anabolic response of osteoblast to low-intensity pulsed ultrasound, J. Bone Miner. Res. 18 (2) (2003) 360–369.

56. K. Hidaka, C. Miyamoto, S. Wada-Takahashi, M. Saita, A. Kawata, R. Kawamata, Y. Maehata, M. Minabe, S. Takahashi, Y. Mikuni-Takagaki, Humoral response to therapeutic low-intensity pulsed ultrasound (LIPUS) treatment of rat maxillary socket after the removal of a molar tooth, Int. J. Anal. Bio-Sci. 3 (1) (2015).

57. K. Bandow, Y. Nishikawa, T. Ohnishi, K. Kakimoto, K. Soejima, S. Iwabuchi, K. Kuroe, T. Matsuguchi, Low-intensity pulsed ultrasound (LIPUS) induces RANKL, MCP-1, and MIP-1beta expression in osteoblasts through the angiotensin II type 1 receptor, J. Cell. Physiol. 211 (2) (2007) 392–398.

58. H. Watabe, T. Furuhama, N. Tani-Ishii, Y. Mikuni-Takagaki, Mechanotransduction activates a5b1 integrin and PI3K/Akt signaling pathways in mandibular osteoblasts, Exp. Cell Res. 317 (18) (2011) 2642– 2649.

59. L. Franchi, R. Munoz-Planillo, G. Nunez, Sensing and reacting to microbes through the inflammasomes, Nat. Immunol. 13 (4) (2012) 325–332.

60. M.D. Schofer, J.E. Block, J. Aigner, A. Schmelz, Improved healing response in delayed unions of the tibia with low-intensity pulsed ultrasound: results of a randomized sham-controlled trial, BMC Musculoskelet. Disorders 11 (2010) 229.

CHAPTER 9

Multidimensional spectral analysis of the ultrasonic radiofrequency signal for characterization of media

Simona Granchi, Enrico Vannacci, Elena Biagi, Leonardo Masotti

HIGHLIGHTS

- A spectral method based on ultrasonic radiofrequency signal is presented in detail.
- High specificity were obtained in density level discrimination.
- High robustness of the algorithm with respect to the signal to noise ratio was tested.
- Conventional ultrasound can be supported in tissue characterization by this method.

ABSTRACT

The importance of the analysis of the radiofrequency signal is by now recognized in the field of tissue characterization via ultrasound. The RF signal contains a wealth of information and structural details that are usually lost in the B-Mode representation. The HyperSPACE (Hyper SPectral Analysis for Characterization in Echography) algorithm presented by the authors in previous papers for clinical applications is based on the radiofrequency ultrasonic signal. The present work describes the method in detail and evaluates its performance in a repeatable and standardized manner, by using two test objects: a commercial test object that simulates the human parenchyma, and a laboratory-made test object consisting of human blood at different dilution values. In particular, the sensitivity and specificity in discriminating different density levels were estimated. In addition, the robustness of the algorithm with respect to the signal-to-noise ratio was also evaluated.

KEYWORDS

Ultrasonic radiofrequency signal; Spectral analysis; Signal processing; Tissue characterization; Quantitative ultrasound

1. INTRODUCTION

When an ultrasonic wave propagates through soft tissue, an interaction occurs between the mechanical energy of the wave and the local structure, generating energy absorption, reflection, and scattering. The energy propagated back toward the ultrasonic transducer constitutes the ultrasonic echo signal called the radiofrequency (RF) signal. The RF signal contains information about ultrasound–tissue interaction [1], [2], [3], [4], [5], [6],[7] and [8] and a processing method must be used that is capable of extracting this information.

The amplitude is related to the distribution of mechanical impedance (density, elastic characteristics) of the backscattering medium, the scatterer concentration and the ratio between the sizes of the microstructure and the wavelength [1], [3], [9] and [10]. The phase information, related to the interferences, depends on the mutual distances and geometrical organization of the tissue microstructure scatterers. These interferences and reflectivity variations in the time domain are responsible for spectral amplitude modulation in the frequency domain. In this context, over the last thirty years quantitative ultrasound (QUS) techniques have been developed [11], [12] and [13] to improve tissue characterization as a support for diagnostics. Indeed, in order to gain further information for tissue characterization and differentiation purposes, it is essential not only to preserve the shape of the RF signal spectrum, but also to identify the spectral parameters that are best correlated with the investigated structures [3], [4], [9], [14], [15], [16], [17], [18], [19],[20], [21], [22], [23], [24] and [25].

Our group used the RF signal for our investigation techniques [26] and [27] and developed the RULES (Radiofrequency Ultrasonic Local EStimators) algorithm [28],[29], [30], [31] and [32] based on the analysis of the local power spectrum obtained by DWPT (Discrete Wavelet Transform). Even though significant results have been achieved in various research fields [23], [27], [28], [29], [33] and [34], they have been below expectations. In fact, the method was dependent on the instrumental parameters of the acquisition setup [24] and [25] and the number of extracted features was insufficient for characterizing tissue with good specificity and sensitivity as demonstrated by the results obtained by comparing the RULES with the method discussed in this paper[35].

The proposed investigation method called HyperSPACE (Hyper SPectral Analysis for Characterization in Echography) is able to extract local information about the tissue under investigation and implements a sub-band spectral decomposition. The method in not directly based on the analysis of the power spectrum of the RF signal, as with RULES and the principal QUS techniques, however it works in a spectral domain of *N*-dimensions; this is the origin of the Hyper suffix in the name, where *N* is the number of sub-bands into which the RF signal bandwidth is decomposed.

The method was applied in an important experimentation involving ten Italian hospital clinics in order to differentiate the two most common breast pathologies [35],[36] and [37]. High values of sensitivity and specificity in differentiating fibroadenoma and infiltrating ductal carcinoma were obtained and compared with histological examinations.

In previous works, we focused our attention on the training phase of the algorithm and the description of the results. In the present paper, the fundamental steps of the algorithm are explained in detail. Moreover, the sensitivity of the method, the dependence of the method on the instrumental parameters, and the robustness of the algorithm with respect to the signal-to-noise ratio (SNR) have been estimated. To do this, it was important to test the method starting from the analysis of a single parameter by using commercial and simple test objects. Two different test objects were used. The first, a commercial CIRS model 047 (Computerized Imaging Reference Systems, Inc. Norfolk, Virginia 23513 USA) test object, consists of background material that simulates the human parenchyma with a series of aggregates at different densities inserted to simulate the presence of cysts. The second was a laboratory test object consisting of a transfusion bag containing human blood at different concentrations. Due to the characteristics of the two test objects, density was chosen as the parameter for evaluating the properties of the HyperSPACE algorithm.

2. INVESTIGATION METHOD

The proposed algorithm can be classified as a QUS technique based on the analysis of the backscattered ultrasonic RF signal. Moreover, it performs a sort of Texture Analysis[13], [38], [39] and [40] as it tries to characterize the echographic image by extracting the typical features that determine the texture. In diagnostic ultrasounds, texture analysis can be performed either directly by analyzing the correlations among spatial gray-levels of B-Mode images, or indirectly with QUS techniques applied to the spectral features. The purpose of the algorithm is to identify a new spectral domain where the signal parameters, correlated with the mechanical characteristics and structural organization of the medium under investigation, can be extracted.

The investigation method works in a hyperspace consisting of N spectral dimensions obtained from a sub-band decomposition of the RF signal in order to read the local spectral amplitude modulations generated by the distribution of the scatterers. Another specific characteristic of the method is the implementation of "local normalization" that takes into account the effective energy of the ultrasonic wave which insonifies the "local" portion of the medium. In addition, the proposed normalization allows the algorithm to be independent from the instrumental acquisition parameters of the echographic scanner, such as Time Gain Compensation (TGC) and transmission power.

The HyperSPACE algorithm procedure is illustrated in Fig. 1. The processing procedure is broken down into four main phases that are explained in detail below.

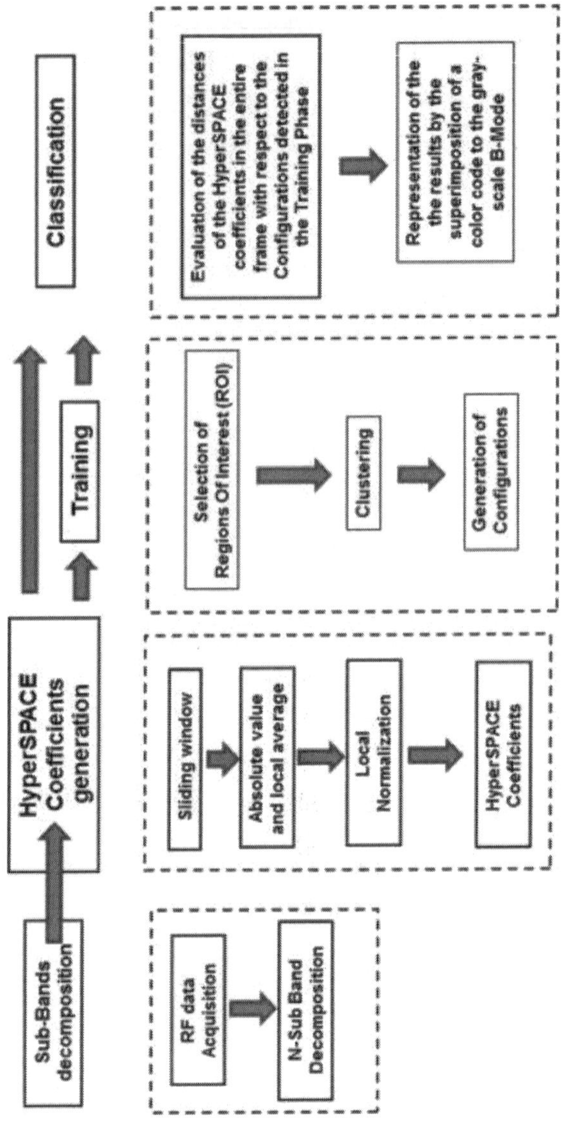

Figure. 1. HyperSPACE (Hyper SPectral Analysis for Characterization in Echography) block diagram. The first step consists of the decomposition of sub-bands for spectral coefficients extraction, the second entails the generation of the HyperSPACE coefficients. The third step is the training procedure on ROI applied on the HyperSPACE coefficients for Configuration generation. The last step is the Classification phase over the entire frame for revealing the presence of Configurations.

2.1. Sub-band decomposition

The first step consists of the decomposition of the signal in the spectral sub-bands, while in the second, the HyperSPACE coefficients are produced. The third is a training procedure applied in order to identify sets of parameters, called Configurations, that are able to characterize the investigated medium. In the fourth phase, the RF frames are classified according to the parameters identified in the previous phase.

The HyperSPACE implements a decomposition of the spectrum by performing a convolution in the time domain between each track of the RF frame with *N* bandpass filters, the impulse responses *h*(*t*) of which have a high degree of correlation with the echographic signal. From among the possible options, the Morlet function, widely used in literature for ultrasonic signals, was chosen as the impulse response [16], [41], [42],[43] and [44]:

$$h(t) = \frac{\beta}{\sqrt{2\pi}} e^{\frac{\beta^2 t^2}{2}} \cos(\mu_k t) \tag{1}$$

In the time domain this signal represents a cosine function modulated by a Gaussian function. The parameters β and μ_k determine the time duration of the response (i.e. its spectral bandwidth) and the central frequency location of the filter respectively. These filters exhibit the same compact support in the time domain. In the spectral domain, they show a different central frequency and the same bandwidth. The bandwidth of the filters and their central frequency spacing depends on the degree of spectral resolution selected.

In this work, the RF signal was acquired with a sampling frequency of 50 MHz and a bank of 24 filters was generated by choosing a constant value of β in order to obtain a bandwidth of 1 MHz at −12 dB, the choice of which is explained in Section 4. The value ofμ_k was set so that the central frequencies of the filters were multiples of 1 MHz. These 24 spectral sub-bands represent the dimensions of the domain in which the HyperSPACE analyzes the signal. The decomposition procedure of the RF signal is illustrated in Fig. 2, where it can be observed how each track of the RF frame (Fig. 2a and b) is decomposed by a filter bank. In Fig. 2c several impulse responses of the filters are shown. If Sb_1, Sb_2, … , Sb_N are the N sub-bands into which the signal spectrum is decomposed, then the RF signal at time t_0 can be represented by N coefficients c_1, c_2, … , c_N, that are the convolution coefficients between the signal and each filter of the bank as shown inFig. 2d. In this way, it is possible to know the corresponding N coefficients in the transformed domain for any instant t_0 of the RF signal. For each RF track formed by Ysamples, Y vectors of length N are obtained and consequently, C_{Sb} matrices with dimensions X by Y are derived for each frame constituted by X tracks. Each of these C_{Sb}matrices contains the coefficients, called $c_{i,m,Sb}$, relating to a particular sub-band, indicated by the subscript Sb; the other two subscripts i and m indicate the position within the matrix (and hence the track and the time instant associated with the coefficient). In our case, 24 C_{Sb} matrices are produced for each RF frame.

2.2. HyperSPACE coefficient generation

In order to increase the performance of a method it is necessary to make it as independent as possible from any amplitude variations determined by the settings of the instrumental acquisition set-up, or by the characteristics of previously encountered structures. For this reason, a distinctive feature of the proposed method consists of performing a normalization process that takes the actual ultrasonic signal into account locally, i.e. within a window of limited size, which is present in the investigated portion of the sample. This operation, which can be called “Local Normalization”, is illustrated inFig. 3, where the entire procedure of the HyperSPACE coefficient generation is shown. The first step (Fig. 3a) consists of a local average performed to reduce the variability of the coefficients. This is carried out by sliding a partially overlapped window on top of the absolute values of each of the N C_{Sb} matrices. The size of the Nmatrices obtained, called

Local Average Matrices (LAM_{Sb}), is K by J depending on the choice of the size of the sliding window and the overlaps, according to (2):

$$K = [(X - Win\_X + 1)/Step\_X]$$
$$J = [(Y - Win\_Y + 1)/Step\_Y] \quad (2)$$

In this work, the window dimensions are Win_X = 6 tracks (1.2 mm), Win_Y = 80 pixels (1.2 mm) and the overlaps Step_X = 2 tracks and Step_Y = 10 pixels.

In (3) the formula that links the coefficients $c_{i,m,Sb}$ to $d_{k,j,Sb}$ is as follows:

$$d_{k,j,Sb} = \frac{1}{Win\_X \cdot Win\_Y} \sum_{i=1}^{Win\_X} \sum_{m=1}^{Win\_Y} |c_{i,m,Sb}| \quad (3)$$

The N LAM_{Sb} matrices are therefore constituted by the coefficients $d_{k,j,Sb}$. In order to assess the magnitude of the ultrasonic signal that locally affects each investigation window considered, the Local Normalization Matrix (LNM) is calculated. In Fig. 3b, it can be seen how the LNM consists of the mean value of the LAM_{Sb} matrices corresponding to the central frequencies of the probe employed. The $e_{k,j}$ coefficients are thus obtained according to the following formula:

$$e_{k,j} = \frac{1}{1 + P_f - P_i} \sum_{Sb=P_i}^{P_f} d_{k,j,Sb} \quad (4)$$

where P_i and P_f are the indices related to the initial and final sub-bands respectively, chosen by considering the central frequency of the transducer used; for the probe LA523 (Esaote S.p.A, Firenze, Italy), $P_i = 4$ and $P_f = 10$.

At this point, each of the LAM_{Sb} is divided by the LNM matrix in order to obtain N new locally normalized matrices (HS_{Sb}) constituted by the $hs_{k,j,Sb}$ coefficients (Fig. 3c) defined by:

$$hs_{k,j,Sb} = \frac{d_{k,j,Sb}}{e_{k,j}} \quad (5)$$

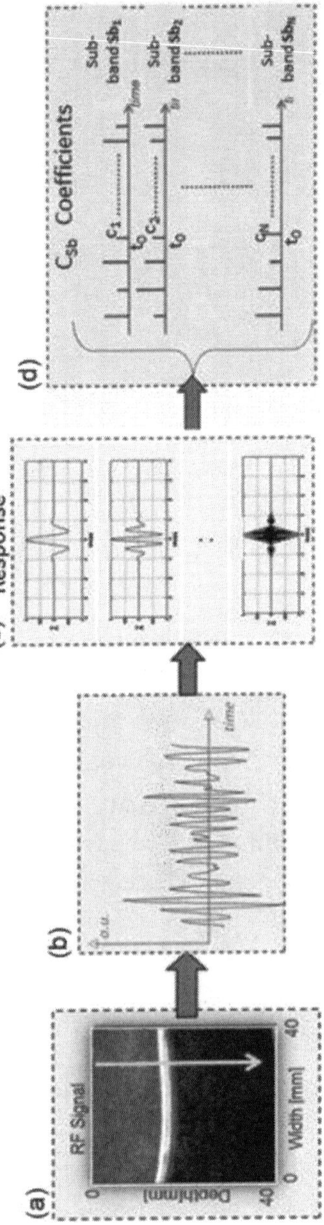

Figure. 2. Sub-Bands decomposition. In (a) the B-Mode of an ultrasonic radiofrequency (RF) frame, related to a transfusion bag of blood is reported. Each track of the RF frame (b) is convolved with *N* filters (c) and Y tracks of spectral coefficients c_1, c_2, … , c_N are produced, as shown in (d). In (c) some impulse responses of the *N* filters are illustrated.

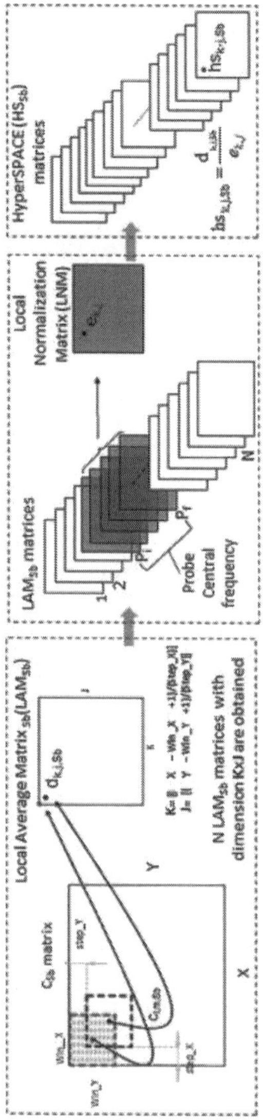

Figure. 3. Local Normalization and Hyperspace Coefficients generation. A local averaging is applied on the absolute values of the N matrices of spectral coefficients (C_{Sb}), in order to obtain Local Average Matrices (LAM_{Sb}) as shown in (a). The Local Normalization Matrix (LNM) is produced by calculating the mean value of the matrices LAM_{Sb} relating to the central frequencies of the employed transducer (b). The HyperSPACE matrices HS_{Sb} are obtained by dividing point to point each matrix LAM_{Sb} by the LNM as illustrated in (c).

These values are the HyperSPACE *hs* coefficients and represent the coordinates of the original samples of the RF frame in the new domain.

At this point, two other phases of the algorithm are carried out, called Training and Classification respectively.

2.3. Training

In the Training phase, Regions of Interest (ROIs) are selected on the B-Mode (Fig. 4a) by the operator. The selection of ROIs is an important operation for ensuring a proper training of the algorithm. In the clinical environment, physicians depending on their experience and based on the echographic appearance draw a closed line around the area to be characterized on the B-Mode. In a laboratory environment, such as in this study, the ROI is selected by taking the echographic appearance and the geometrical, structural and physiological characteristics provided by manufacturers of the test objects into account.

Subsequently, the $hs_{k,j,\mathrm{Sb}}$ coefficients relating to the selected ROIs are analyzed. Fig. 4a shows the B-Mode of a portion of a commercial CIRS model 047 test object (described in detail in Section 3) containing three aggregates with densities differing from the background material they are immersed in. From left to right, the aggregates with +3 dB, +6 dB, and +9 dB are selected respectively on the B-Mode with three different ROIs: ROI 1, ROI 2, and ROI 3. The coefficients of the three ROIs relating to areas with different physical characteristics have different shapes and positions in the hyperspace considered. In order to characterize their distributions, the coefficients are therefore subjected to a clustering process through the iterative algorithm K-means [45]. By considering a partition of data into N clusters, the K-means algorithm is able to define the position of the centroids through an iterative procedure that minimizes the total point-to-centroid distances summed over all N clusters. After this, each cluster is identified by its centroid and by a distance function able to evaluate the degree of separation between clusters.

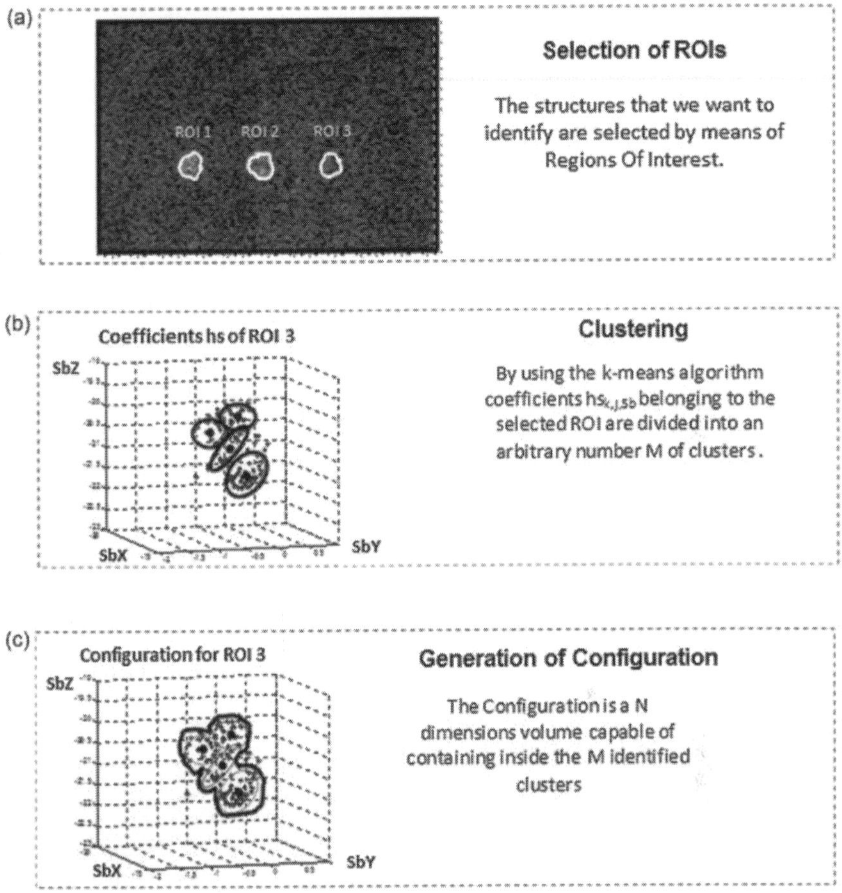

Figure. 4. Training Phase procedure. In (a) Regions Of Interest (ROIs) are identified in the B-Mode image in order to select their *hs* coefficients. The Coefficients are clustered by K-means algorithm and M clusters are generated for each selected ROI, as shown in (a). In (b) the clusters of ROI 3 are considered. The *N*-dimensional volume which collects the M clusters in each ROI is called Configuration, as visualized in (c) for the ROI 3.

Fig. 4b shows the different clusters into which the $hs_{k,j,\mathrm{Sb}}$ coefficients related to the ROI 3 are divided. The combination of all the clusters relating to each ROI identifies a hyper volume in the hyperspace characterized by the M centroid and the M by N distances of the clusters it consists of, and which will be called a Configuration (Fig. 4c). This hyper-volume can also be synthetically localized by an average centroid. This is obtained by computing the arithmetic mean of the M centroids belonging to the Configuration.

2.4. Classification

In the Classification phase the $hs_{k,j,\mathrm{Sb}}$ coefficients relating to the whole frame are analyzed and it is determined whether each coefficient belongs to one of the Configurations developed during the Training or not. For each $hs_{k,j,\mathrm{Sb}}$ coefficient, the distance from the centroid of the considered Configuration is calculated in each dimension. If the coefficient is found inside the hyper volume defined by the Configuration, this coefficient will be considered as belonging to the latter. The visualization of the final result is carried out by associating a color code to the pixels of the image related to the $hs_{k,j,\mathrm{Sb}}$ coefficients that belong to each Configuration. This color is then superimposed on the B-Mode grayscale image (Fig. 5). Therefore, if for example three Configurations are present in a frame, the final B-Mode will have three overlapping color codes corresponding to the areas identified by the three Configurations. Fig. 5a illustrates the B-Mode relating to a section of the commercial test object described above. Fig. 5b contains the final results of the Classification. The three Configurations are shown on the B-Mode in red, green and blue, allowing easy identification of the three aggregates with different densities.

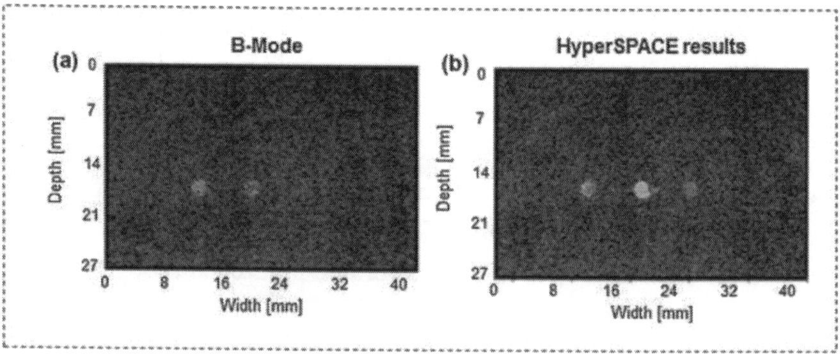

Figure. 5. The Classification phase is performed on the entire analyzed frame in order to detect the presence of the identified Configurations. The figure shows the results obtained on CIRS model 047 (Computerized Imaging Reference Systems, Inc. Norfolk, Virginia 23513 USA) tissue mimicking phantom. The phantom presents cysts with a density differing from the background material. (a) B-Mode image. (b) HyperSPACE results. Each Configuration is associated to a different color. (For interpretation of the references to color in this figure legend, the reader is referred to the web version of this article.)

3. MATERIALS AND METHODS

The algorithm was tested using calibrated test objects in order to evaluate the ability of the method to detect density variations. Two different test objects with variable densities were used.

The first, as mentioned above, was a commercial CIRS model 047 test object consisting of a background material that simulates the human parenchyma with a series of aggregates with different densities (from −9 dB to +9 dB) and with diameters of 2.4 mm, 4 mm and 6.4 mm, inserted to simulate the presence of cysts. The second was a laboratory test object consisting of a transfusion bag containing human blood immersed in a tank filled with degassed water.

The laboratory tests were performed according to the guidelines of the University of Florence, on human blood provided by a healthy donor who had previously signed an informed consent form. The transfusion bag was filled with different blood concentrations obtained from multiple dilutions with physiological solution. Bags with a capacity of 250 ml and the anticoagulant CPDA (Citrate – Phosphate – Dextrose – Adenine) were used. Through a calibrated and repeatable procedure, 31 different values of Fractional Dilution (FD) were obtained. The FD was defined as follows:

FD = $Vol_{blood}/Vol_{blood+physiological\ solution}$, where Vol_{blood} is the quantity of blood introduced into the bag and $Vol_{blood+physiological\ solution}$ corresponds to the entire volume of the liquid contained in the same bag. In Fig. 6, the sequence of FD values for the blood samples is reported in a logarithmic diagram.

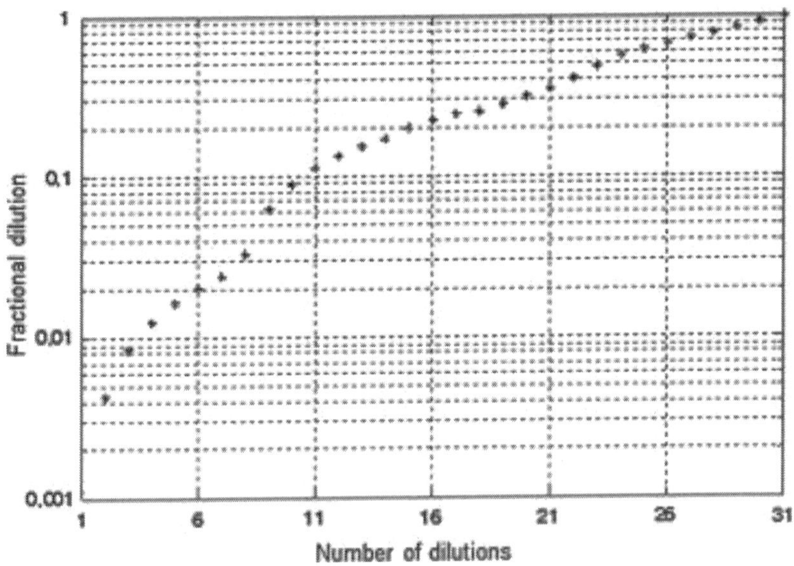

Figure. 6. Values of Fractional Dilutions (FD) of the all blood analyzed samples.

Fifty RF frames were acquired and processed for each concentration in order to assess the parameter variations as a function of the FD. This second test object was selected to allow use of a simple and repeatable biological tissue with a variable distribution of scatterers.

In fact, as widely discussed in literature [46], [47], [48] and [49], blood can be described as a medium [47], [48] and [50] composed of aggregates of red blood cells immersed in plasma. The back-propagated RF signal depends on the concentration of the scatterers and other parameters related to the blood flow. The relation between the frequency of the RF signal, the concentration, and the blood flow parameters has been extensively studied and tested by Cloutier et al. [47], [48] and [51]. Fig. 7 contains the acquisition system. All RF data were collected by means of the FEMMINA (Fast Echographic Multi parameters Multi Image Novel Apparatus) [52] and [53], a hardware-software platform developed by the Ultrasound and Non-Destructive Testing Laboratory. FEMMINA was used to perform multi-parameter extractions and multi-image data representations starting from the acquired RF signal. For this experiment, FEMMINA was connected to a commercial ultrasound scanner (Mylab 90, Esaote S.p.A, Firenze, Italy), as shown in Fig. 6, which was modified to provide the RF signal output. The ultrasound scanner provides RF beam-formed data sampled at 50 MHz and amplified by the Time Gain Compensation (TGC) module. The data are transferred to FEMMINA by means of a 1 Gbit/s optical fiber connection. FEMMINA acquires RF data in real time mode at the same frame rate as the scanner. RF data frames can also be stored in the PC memory and then saved on the hard disk to enable an off-line frame-by-frame evaluation. The ultrasonic probe used for this experimentation was an LA523 linear array with a nominal frequency of 7.5 MHz.

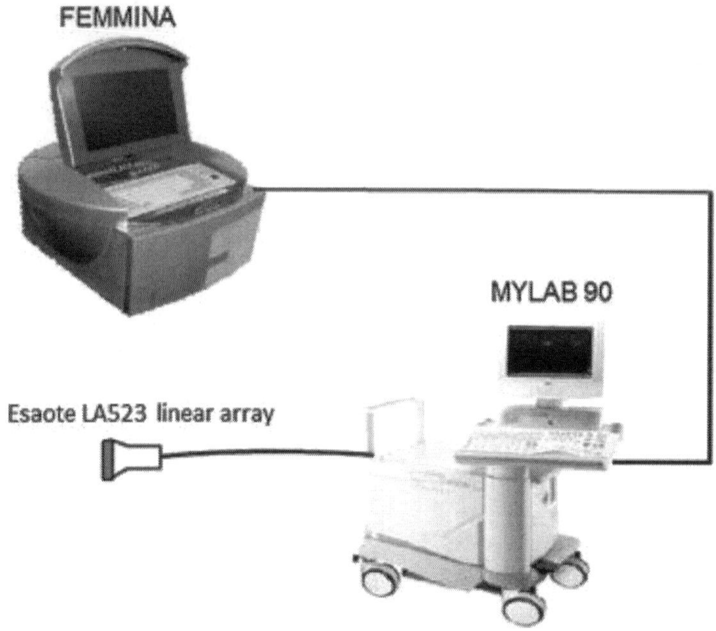

Figure. 7. Acquisition set-up. FEMMINA is connected to the commercial ultrasound scanner Mylab 90 (Esaote S.p.A, Firenze, Italy). The ultrasonic probe is an LA523 linear array (Esaote S.p.A, Firenze, Italy) with a nominal frequency of 7.5 MHz.

4. EXPERIMENTAL RESULTS

The algorithm was tested on the CIRS model 047 test object in order to achieve three targets:

1) to evaluate the possibility of differentiating the density of hyper echoic cysts with respect to the background material;

2) to verify whether HyperSPACE is able to reduce attenuation effects, allowing the same Configuration to better identify cysts with the same density at different depths, also thanks to the Local Normalization procedure described in Section 2 and illustrated in Fig. 3;

3) to observe how the Local Normalization makes it possible to detect the same Configuration for the same cyst by varying the ultrasound scanner acquisition settings such as focus depth, transmission power, and total gain.

Before carrying out these experiments, several preliminary tests were performed in order to select the best frequency range subdivision. Different sub-band widths of filters were tested: 3 MHz, 1.5 MHz, 1 MHz and RF frames referring to proximal cysts of the CIRS model 047 were processed. The goodness of the processed images was evaluated in terms of ability to discriminate the density levels and differentiate the cysts from the surroundings. The best compromise was obtained using a 1 MHz frequency sub-band width, as can be observed in Fig. 8 which contains processed HyperSPACE images using 3 MHz, 1.5 MHz and 1 MHz sub-band widths. To achieve the first two objectives, 10 RF frames related to sections not spatially correlated with each other were acquired by moving the probe on the CIRS model 047. The acquired sections contain two rows of cysts with diameters of 2.4 mm and 4 mm respectively at different depths. The acquisition settings were the following: a transmission power of 80%, TGC of 50% at each depth, a total gain of 55%, and a focus position at 20 mm. One of these RF frames was used to train the algorithm and three different ROIs, one for each cyst proximal to the transducer, were selected as illustrated in Fig. 9a. As mentioned above in Section 2, the selection of ROIs is important for ensuring proper training of the algorithm and for this reason we selected the three cysts depending as much as possible on their size. The other 9 sections were used in the Classification phase. The Training phase allowed for identifying the three Configurations, each representing a different degree of density. The Configurations are colored in red, green and blue for the +3 dB, +6 dB, and +9 dB densities respectively. In Fig. 9b and c, it is possible to see how cysts with the same density are characterized by the same Configurations. In particular, it is important to observe how the algorithm detects the presence of the aggregates with a density of +3 dB, which is barely distinguishable from the background material in the B-Mode image. It is even more interesting to note how the cysts with the same density are characterized by the same Configurations, regardless of their different depths, thanks also to the Local Normalization process which reduces the attenuation effects.

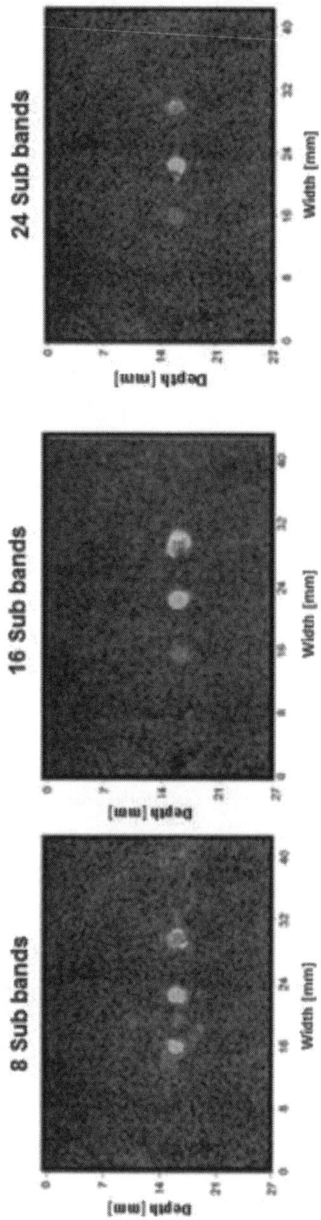

Figure. 8. Three processed HyperSPACE images obtained by decomposing the signal spectrum by means of filters banks with different bandwidths. The first, on the left, is related to a decomposition bank of 8 filters with bandwidth = 3 MHz, the second to a bank of 16 filters with bandwidth = 1.5 MHz and the last to a bank of 24 filters with bandwidth = 1 MHz.

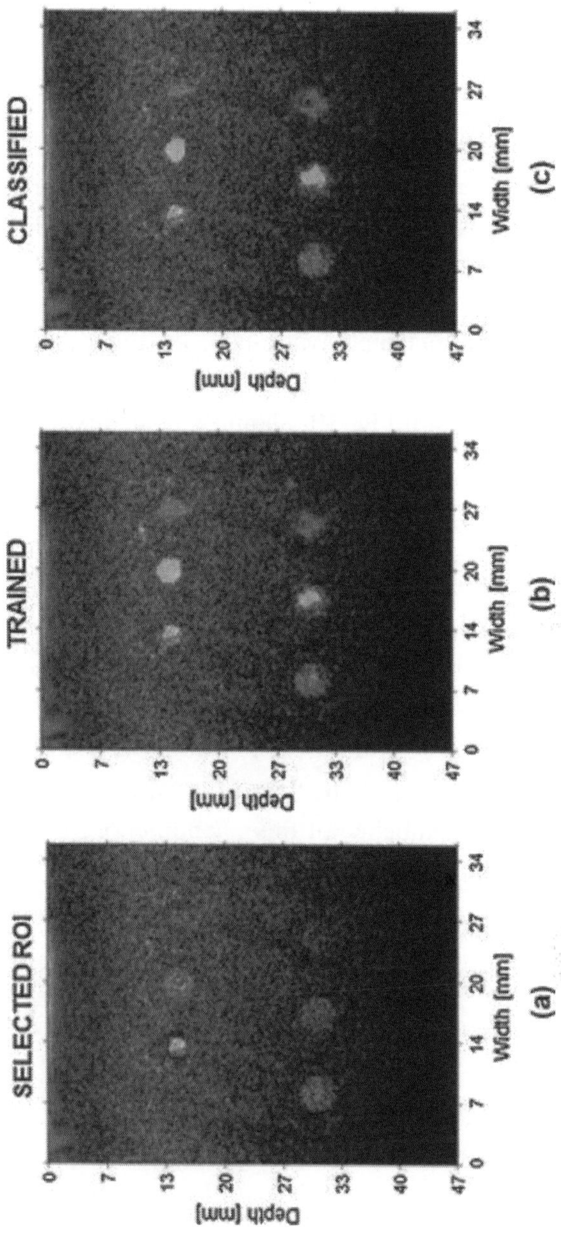

Figure. 9. Training and Classification phase results for CIRS test object. In (a) the selected ROIs are shown. In the Training, only proximal cysts are considered for setting the Configurations. (b) and (c), show the results of the processed frame in Training and Classification, after normalization procedure.

In order to test the Configurations by changing the acquisition settings, three other measurement campaigns were performed. For each one, 7 RF frames relating to the same section of the CIRS model 047 were acquired by modifying only one parameter of the ultrasound scanner at each acquisition. Each of the 7 acquired frames was classified by means of the Configurations identified in the previous training phase, as shown in Fig. 9.

In the first set, the focus depth was changed from a distance of 52 mm to 10 mm. The results relating to the focus depths of 52 mm, 20 mm and 15 mm are shown inFig. 10a, b and c respectively. In the second set, the parameter considered was the total gain, the dynamics of which were increased from 55% to 90%. The frames with a gain of over 65% were not classified due to the spectral modification effects caused by the saturation of the acquisition system electronics. The results obtained for gains of 55%, 60% and 65% are shown in Fig. 10d, e and f respectively. Gains greater than 65% cause signal saturation and thus the HyperSPACE, which performs a spectral analysis, cannot be applied because it would produce results affected by errors. The incorrect classification of several regions in the images 10e and 10f is a consequence of this phenomenon. In fact, the colors representing the three Configurations are also in the background surrounding the cysts.

The last set was related to the variations in the transmission power parameter within a dynamics range of from 25% to 100%. Figs. 10g–i show the results obtained from the processing of three frames acquired at transmission powers of 80%, 60%, and 40% respectively. As can be seen from all three acquisition campaigns, the algorithm demonstrated good stability due to the three density levels being properly characterized for all the classified frames, irrespective of the variations in the acquisition settings considered.

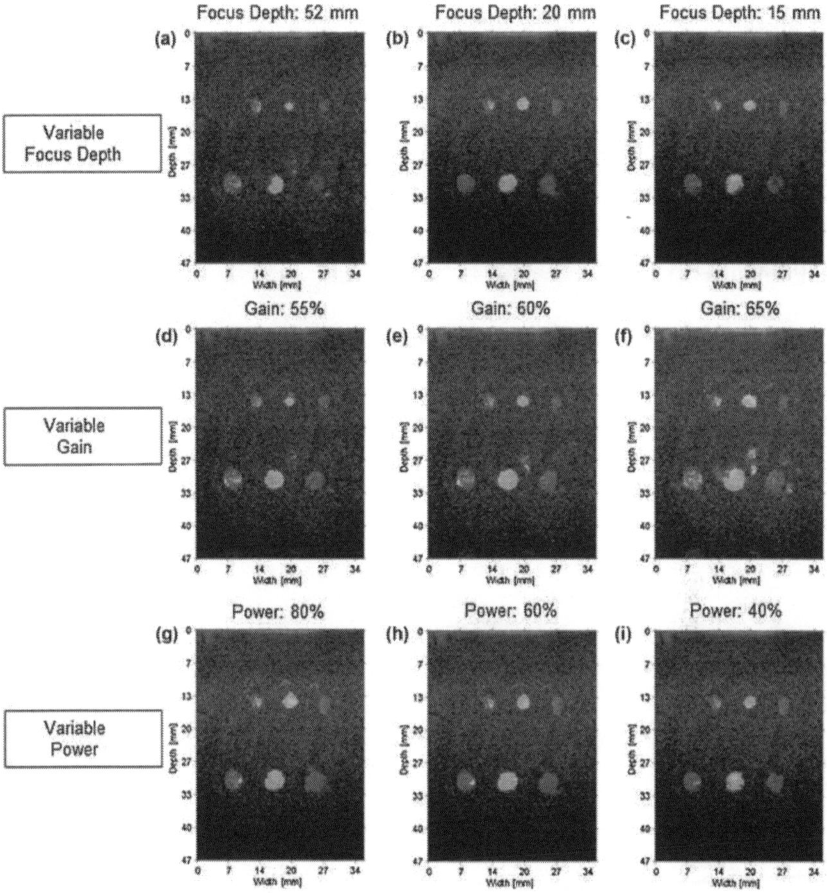

Figure. 10. In (a–c) Classification results obtained at variable focus depths. In (d–f) Classification results obtained for different gains and in (g–i) Classification results obtained for different transmission powers. The analyzed sections contain two rows of cysts with diameters of 2.4 mm (proximal) and 4 mm (at a greater depth).

The algorithm was then tested on the test object consisting of human blood. Fig. 11shows some of the dilutions with different FDs represented in one of the possible two-dimensional projections of the hyperspace, corresponding to SbX = 3 MHz and Sby = 13 MHz. These two sub-bands have been chosen by way of example. It can be observed how the distribution of the *hs* coefficients is influenced by changing the FD. This induced us to consider the possibility of identifying different concentration levels with specific Configurations. In fact, the clusters of the coefficients labeled FD = 0, which refers to the physiological solution, and FD = 0.005, FD = 0.01, FD = 0.48 and FD = 1.00 corresponding to the human blood, are easily distinguished in this specific two-dimensional projection of the hyperspace.

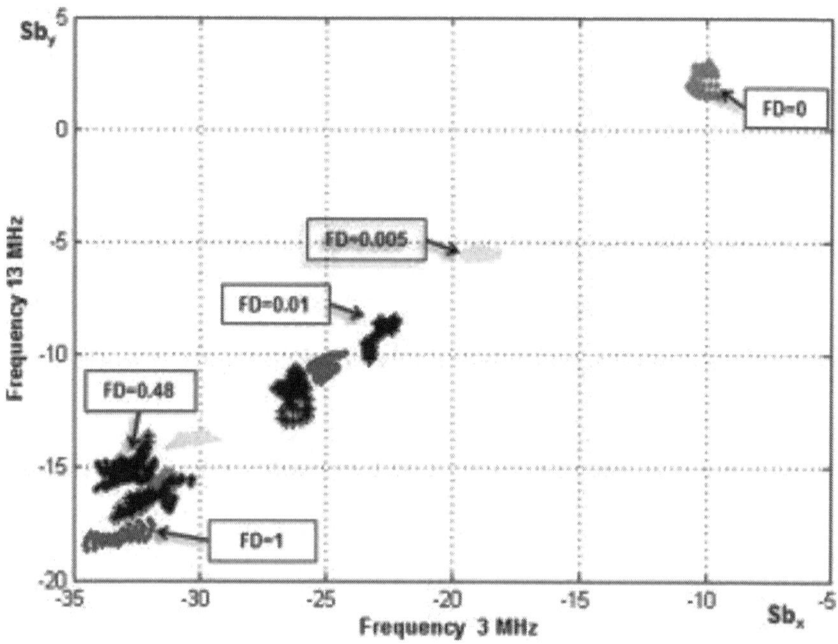

Figure. 11. Example of coefficients *hs* at different FD values is shown in a two-dimensional projection of the *N*dimension hyperspace. The choice of these sub-bands in the figure is only to give an example. It is worth noting that different clusters correspond to different FD values.

The use of this test object made it possible to test the algorithm performances. First of all, the capability of producing a Configuration to detect low concentration levels in order to evaluate the electronic sensitivity of the method and moreover the ability to set two different Configurations for acquisitions with FDs very close to each other and with very similar echogenicity in the B-Mode image. For these reasons, FD = 0.01, FD = 0.48 and FD = 0.68 were chosen for training the algorithm. The other frames corresponding to different FD values were only used in the Classification phase. The training produced three Configurations: the first is shown in red relates to FD = 0.68, the second in green to FD = 0.48 and the last in blue to FD = 0.01.

The average values of the gray levels calculated within a window inside the blood bag for each of the 31 dilutions are reported in Fig. 12. It is worth noting, for example, how the frames with FD = 0.48 and FD = 0.68 show very similar gray-level values. Fig. 13contains the colored area percentages relating to the three Configurations produced in the training phase. The percentages were estimated on each of the 31 frames, with respect to the entire frame in order to take into account the capability of each Configuration both to characterize only one specific FD value and not to mark areas outside the bag blood, as represented in the B-Mode of Fig. 12 and Fig. 14 by the regions below the strong reflection. A good degree of selectivity for the Configurations relating to FD = 0.48 (green), and FD = 0.68 (red) can be observed, as they are minimally superimposed. Fig. 14 contains nine processed images relating to different dilutions. In addition to the three frames used in the training phase, several frames subjected to the classification phase only are also reported. Also in this case, as confirmed by Fig. 13, it is possible to observe how the three Configurations seem to be very selective, i.e. the colored areas on the classified frames are either very small or absent. In Fig. 14 it is evident that the two green and red Configurations, in addition to differing one from the other, do not appear in the frames with dilutions close to FD = 0.48 or FD = 0.68. It can also be observed how the frames used for the training phase are not uniformly colored, due to the inevitable formation of aggregation centers [47].

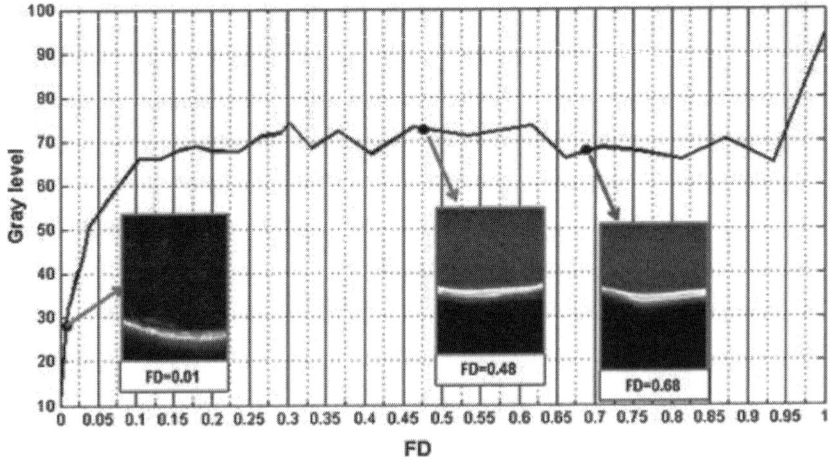

Figure. 12. Gray level mean value calculated on the B-Mode images, in relation to different FD values. For FD from 0.25 up to 0.9, the gray level mean has small fluctuations and therefore it is very difficult to distinguish the different dilution levels through the analysis of the B-Mode images.

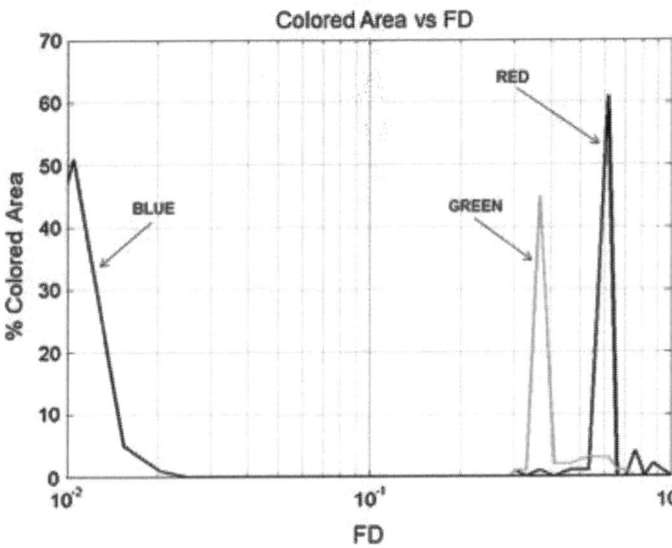

Figure. 13. Percentage of Colored Area calculated on every frame with different FD depending on the FD values. The percentage is calculated for the entire frame. It can be noted how the three colors have very low superimpositions for different FD values and are very specific for the FD selected for the training.

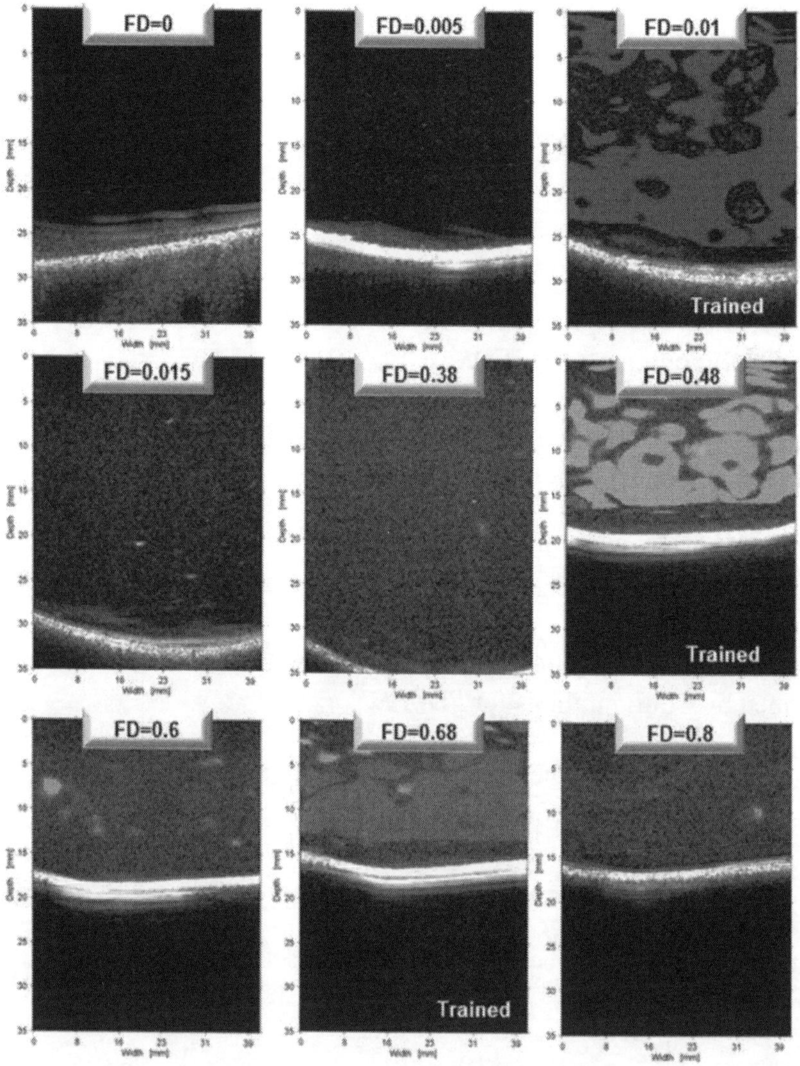

Figure. 14. Results obtained on blood bags with different FD values are reported. Three Configurations have been set: the blue for FD = 0.01, the green for FD = 0.48 and the red for FD = 0.68. They are very specific, that is, they mainly characterize the frames of the blood bags with these FD specific values of FD and only very small colored spots are visible for the other FD values. The algorithm also shows a high electronic sensitivity, in fact, it is capable of detecting the presence of blood for very small concentrations (FD = 0.01). (For interpretation of the references to color in this figure legend, the reader is referred to the web version of this article.)

In order to calculate the specificity (Sp_{FDK}) and sensitivity (Se_{FDK}) of each Configuration, the following expressions were applied [54]:

$$Se_{FDK} = \frac{TP_{FDK}}{TP_{FDK} + FN_{FDK}} \qquad Sp_{FDK} = \frac{TN_{FDK}}{TN_{FDK} + FP_{FDK}} \tag{6}$$

where

TP_{FDK} : area of regions into blood bags with FD=*FDK* truly stained by Configuration for FD=*FDK*;

FN_{FDK}: area of regions into blood bags with FD=*FDK* falsely not stained by Configurations for FD=*FDK*;

FP_{FDK} : area of regions into blood bags with $FD \neq FDK$ falsely stained by Configuration for FD=*FDK*;

TN_{FDK}: area of regions into blood bags with $FD \neq FDK$ truly not stained by Configurations for FD=*FDK*.

Estimated values for the defined Configurations are reported in Table 1.

Table 1. Values of specificity (Sp_{FDK}) and sensitivity (Se_{FDK}) of each Configuration.

Configuration	Fractional dilution (FDK)	Se_{FDK} (%)	Sp_{FDK} (%)
Red	0.68	71.0	99.6
Green	0.48	50.2	99.6
Blue	0.01	44.5	99.8

The obtained values are due to the choice of having trained the method favoring the specificity with respect to sensitivity. Actually, in many cases the clinical diagnosis is concerned, not only to detect the presence of disease, but especially to differentiate its nature from other types of pathologies with high level of safety. Another consideration needs to be made about Se_{FDK}, that, according to the definition of TP_{FDK} and FN_{FDK}, resulted to be affected by the non homogeneous FD distribution into the blood bags, as said above.

Moreover, the minimum FD level detectable by the BLUE Configuration was FD = 0.01 and represent the electronic sensitivity of HyperSPACE method. In fact, it was decided to set a Configuration for FD = 0.005, however, FD = 0 was also marked in the Classification, demonstrating that noise and no blood was detected.

It must be pointed out that to assess the stability of a processing method, it is necessary to evaluate its performance as a function of the signal-to-noise (SNR) ratio. In order to achieve this for the HyperSPACE, the acquisition with FD equal to 1 is considered. Moreover, an acquisition was performed with the transducer immersed in water with identical parameters to those used in the acquisition with FD = 1. Different values of synthetic coherent noise created in the Matlab environment (MathWorks, Natick, Massachusetts, USA) were added to this acquisition to obtain 32 different frames with increasing noise. The synthetic noise was filtered by the frequency response of the probe used. On each of these frames, the same ROI was selected in the focalized region and the noise was calculated for the 32 different values of the added synthetic noise, according to the formula:

$$\mathrm{SNR} = 10 * \log 10 \frac{((Signal + Electronic\ Noise)_{FD1})^2}{(Electronic\ Noise_{WATER} + Synthetic\ Noise)^2} \quad (7)$$

where

(Signal + Electronic Noise)$_{FD}$: acquired signal for FD = 1;

Electronic Noise$_{WATER}$: acquired signal in water.

Fig. 15a contains the frame for FD = 1 with the selected ROI. In this region the SNR is equal to 28 dB, as derived from formula (7) with no added synthetic noise. The FD = 1 RF frame without any synthetic noise was processed in order to define a specific Configuration relating to the corresponding concentration level. After which, each frame corresponding to a different value of SNR was processed. The consistency of the colored-area Configuration was tested by evaluating the area percentage with respect to the entire ROI area for each different level of added noise.

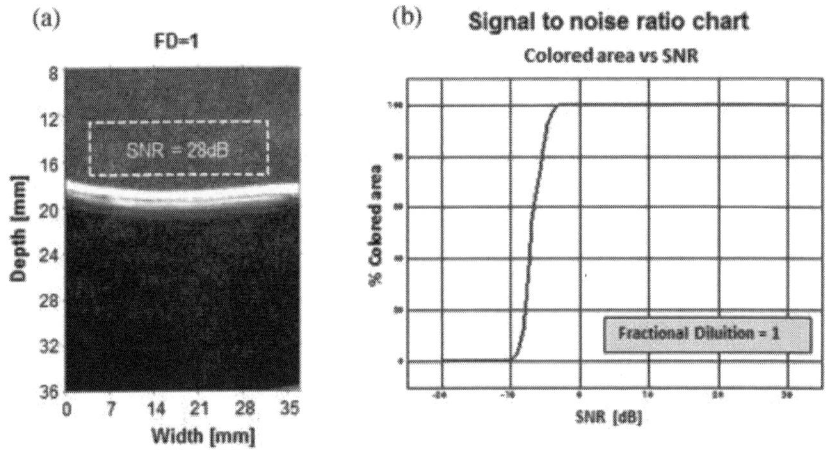

Figure. 15. The acquisition with FD equal to 1.0 was gradually added with synthetic additive noise. A ROI was selected in all the processed frames and its percentage of colored area was evaluated. (a) Selected ROI. (b) Resulting curve at different levels of SNR.

Fig. 15b contains the graph representing the percentages of covered area as a function of SNR.

It can be observed how the percentage of the colored area remains unchanged at 100% of the ROI until the SNR level reaches 3 dB.

When the noise increases further, the percentage of colored area decreases in an almost linear manner until finally disappearing when the SNR level is equal to −10 dB.

5. DISCUSSION

Over the last three decades, researchers using QUS techniques have demonstrated that the interaction between the ultrasonic signal and the investigated structures gives rise to alterations in the RF amplitude and frequency. As already affirmed by the authors in previous publications, the HyperSPACE algorithm extracts spectral features correlated to the structures of the investigated medium. More specifically, the medium density was analyzed in this work by using two stable and repeatable test objects. The use of these kinds of test objects allowed for the standardization and repeatability of the procedures. With the commercial CIRS model 047, it was possible to change the analyzed parameter with different resolutions through different dilution levels in the blood bags, as well as ensure the replicability of results.

The study of spectral parameters enables detection of RF signal alterations due to the organizational and structural characteristics. The analysis of the results highlights how spectral parameters extracted via use of the proposed method provide a significant increase in information regarding the investigated structure in comparison with that supplied by the B-Mode. For example, from the analysis of the B-Mode of the CIRS phantom in Fig. 5, the +3 dB cysts are difficult to distinguish; in addition, from the observation of the B-Modes related to blood bags as shown in Fig. 12, it is very difficult to discriminate the levels of dilution from FD = 0.25 up to FD = 0.9. The HyperSPACE exploits a different type of processing that analyzes the signal in a new spectral domain in which it appears possible to obtain more information about the structure and the organization of the investigated medium. In fact, during the Training phase, the proposed method is able to characterize the three cysts with three different densities by providing a Configuration for each density value. In the Classification, the cysts are correctly detected by the same Configurations without marking external areas even in uncorrelated sections or for different acquisition settings of the ultrasound scanner, as illustrated in Fig. 10.

In addition, the analysis of blood bags has demonstrated that it is possible to define a Configuration for every FD level with high specificity. The clusters referring to different FD values in fact have different shapes and positions in the hyperspace as confirmed in Fig. 11. In the present experimentation, only three Configurations were set and discussed for the sake of simplicity.

Indeed, the "local normalization" of the coefficients is a specific and innovative characteristic of the proposed method as it enables normalization of the spectral contents of each "local" portion of the investigated medium from the energy of the ultrasonic wave that has actually insonified it.

The results obtained from the analysis of different SNR levels have demonstrated the stability of the colored area, confirming the validity of the selected Configuration. By gradually inserting the synthetic additive noise, the percentage of the colored area remains unchanged at 100% of the ROI until the SNR level reaches 3 dB. When the noise further increases, the percentage of colored area decreases in an almost linear manner until finally disappearing when the level of SNR is equal to −10 dB. This fact is very encouraging for future applications in tissue characterization where low SNR conditions occur as in the case of micro vascularization detection.

6. CONCLUSION

This paper presents the fundamental stages and properties of the HyperSPACE analysis method. The algorithm has already been applied to *in vivo* experimentation on breast tissue differentiation where it has achieved significant results. Based on RF ultrasonic signal processing, it works in a *N*-dimensional spectral domain. In the Training phase, the method characterizes areas at different densities by selecting ROIs with different structural organizations and producing Configurations. In the Classification phase, where the entire frame is processed, the aim of the method is to detect the regions of the investigated medium the hyperspace *hs* coefficients of which belong to previously identified Configurations. Regions with the same density are correctly detected by the specific Configurations without marking any other portions. The ability of the HyperSPACE to differentiate density levels has been evaluated via use of two stable and repeatable test objects.

The two specific characteristics of the method are the identification of a new spectral domain, the dimension of which depends on the number of sub-bands into which the RF signal is decomposed, and the "local normalization" that allows for analyzing variations in the back-propagated signal compared to the actual ultrasonic signal present in the investigated portion of the medium. Moreover, the HyperSPACE has demonstrated great robustness with respect to the signal-to-noise ratio. In fact, the method was able to recognize the Configuration with a 50% covered area at a SNR equal to −7 dB.

ACKNOWLEDGMENT

This work was promoted and funded by the Regione Toscana within the FORTE research project of the operational regional program Por Creo Fesr 2007–2013.

REFERENCES

1. G. Ghoshal, J. Mamou, M.L. Oelze, State of art methods for estimating backscater coefficient, in: J. Mamou, M.L. Oelze (Eds.), Quantitative Ultrasound in Soft Tissues, Springer, 2013, pp. 3–20 (Chapter 2).

2. E.J. Feleppa, M.M. Yaremko, Ultrasonic tissue characterization for diagnosis and monitoring, IEEE Eng. Med. Biol. 6 (4) (1987) 18–26.

3. E.J. Feleppa, L. Liu, A. Kalisz, M.C. Shao, N. Fleshner, V. Reuter, W.R. Fair, Ultrasonic spectral-parameter imaging of the prostate, Int. J. Imag. Syst. Technol. 8 (1997) 11–25.

4. E.J. Feleppa, J.A. Ketterling, A. Kalisz, S. Urban, C.R. Porter, J.W. Gillespie, P.B. Schiff, R.D. Ennis, C.S. Wuu, W.R. Fair, Advanced ultrasonic tissue typing, and imaging based on radio-frequency spectrum analysis, and neural-network classification for guidance of therapy, and biopsy procedures, in: Proc CARS, vol. 15, 2001, pp. 333–337.

5. R. Jurkonis, A. Janušauskas, V. Marozas, D. Jegelevicˇius, S. Daukantas, M. Patašius, A. Paunksnis, A. Lukoševicˇius, Algorithms and results of eye tissues differentiation based on RF ultrasound, Sci. World J. (2012) 1–6.

6. M. Moradi, P. Abolmaesumi, P. Mousavi, Tissue typing using ultrasound RF time series: experiments with animal tissue samples, Med. Phys. 37 (2010) 4401–4413.

7. G. Schmitz, H. Ermert, T. Senge, Tissue-characterization of the prostate using radio frequency ultrasonic signals, IEEE Trans. Ultrason., Ferroelectr., Freq. Control 46 (1999) 126–138.

8. R.F. Wagner, M.F. Insana, D.G. Brown, Statistical properties of radio-frequency and envelope-detected signals with applications to medical ultrasound, J. Opt. Soc. Am. A 4 (5) (1987) 910–922.

9. F.L. Lizzi, M. Astor, E.J. Feleppa, M. Shao, A. Kalisz, Statistical frame work for ultrasonic spectral parameter imaging, Ultras. Med. Biol. 23 (1997) 1371– 1382.

10. M. Pereyra, H. Batatia, Modeling ultrasound echoes in skin tissues using symmetric a – stable process, IEEE Trans. Ultrason., Ferroelectr., Freq. Control. 59 (1) (2012) 60–72.

11. G. Ghoshal, M.L. Oelze, W.D. O'Brien Jr., Quantitative ultrasound history and successes, in: J. Mamou, M.L. Oelze (Eds.), Quantitative Ultrasound in Soft Tissues, Springer, 2013, pp. 21–42.

12. J. Mamouy, E. Saegusa-Beecroft, A. Coronzx, M.L. Oelze, L.Yamaguchi, M. Hata, E. Yanagihara, J. Machi, P. Laugierzx, E.J. Feleppay, F.L Lizzi, Spatial-resolution optimization of 3D high-frequency quantitative ultrasound methods to detect metastatic regions in human lymph nodes, in: IEEE Int. Ultrason. Symp., 2013, pp. 1216–1219.

13. H. Tadayyon, A. Sadeghi-Naini, G.J. Czarnota, Noninvasive characterization of locally advanced breast cancer using textural analysis of quantitative ultrasound parametric images, Translational Oncol. 7 (2014) 759–767.

14. M. Aboofazeli, P. Abolmaesumi, G. Fichtinger, P. Mousavi, Tissue characterization using multiscale products of wavelet transform of ultrasound radio frequency echoes, in: Proc. of 31st Ann. Int. Conf. of the IEEE EMBS, Minneapolis, Minnesota, USA, September 2–6, 2009, pp. 479–482.

15. K.D. Donohue, L. Huang, T. Burks, F. Forsberg, C.W. Piccoli, Tissue classification with generalized spectrum parameters, Ultras. Med. Biol. 27 (2001) 1505– 1514.

16. G. Georgiou, F.S. Cohen, Tissue characterization using the continuous wavelet transform Part I: decomposition method,

IEEE Trans. Ultrason., Ferroelectr., Freq. Control 48 (2) (2001) 355–363.

17. P. Guillemain, R. Kronland-Martinet, Characterization of acoustic signals through continuous linear time-frequency representations, in: Proc. IEEE, vol. 84, no. 4, 1996, pp. 561–587.

18. F.L. Lizzi, M.A. Laviola, D.J. Coleman, Tissue signature characterization utilizing frequency domain analysis, in: Proc. IEEE Ultrasonics Symp., 1976, pp. 714– 719.

19. F.L. Lizzi, M. Greenebaum, E.J. Feleppa, M. Elbaum, D.J. Coleman, Theoretical framework for spectrum analysis in ultrasonic tissue characterization, J. Acoust. Soc. Am. 73 (4) (1983) 1366–1373.

20. F.L. Lizzi, E.J. Feleppa, S.K. Alam, C.X. Deng, Ultrasonic spectrum analysis for tissue evaluation, Pattern Recogn. Lett. 24 (2003) 637–658.

21. U. Scheipers, H. Ermert, H.J. Sommerfeld, M. Garcia-Schürmann, T. Senge, S. Philippou, Ultrasonic multifeature tissue characterization for prostate diagnostics, Ultras. Med. Biol. 29 (8) (2003) 1137–1149.

22. G. Schmitz, H. Ermert, T. Senge, Tissue characterization of the prostate using Kohonen-maps, in: Proc. IEEE Ultrasonics Symp., vol. 2, 1994, pp. 1487–1490.

23. J. Wang, C. Kang, X. Liu, T. Li, Y. Wang, T. Feng, Z. Li, N. Xue, K. Shi, Clinical value of radiofrequency ultrasonic local estimators in classifying breast lesions, J. Ultras. Med. 32 (1) (2013) 83–92.

24. J. Wang, C. Kang, T. Feng, J. Xue, K. Shi, T. Li, X. Liu, Y. Wang, Effect of instrument settings on liquid-containing lesion images characterized by radiofrequency ultrasound local estimators, Z. Med. Phys. 23 (2) (2013) 94– 101.

25. J. Wang, C. Kang, T. Feng, J. Xue, K. Shi, T. Li, X. Liu, Y. Wang, Effects of instrument settings on radiofrequency ultrasound local estimator images: a preliminary study in a gallbladder model, Mol. Med. Rep. 8 (4) (2013) 995– 998.

26. E. Biagi, L. Masotti, L. Breschi, M. Calzolai, L. Capineri, S. Granchi, M. Scabia, Radiofrequency real time processing: ultrasonic spectral images and vector doppler investigation, in: Proc. 25th Int. Symp. Acoust. Imaging, vol. 25, 2000, pp. 419–426.

27. L. Masotti, E. Biagi, A. Acquafresca, L. Breschi, M. Calzolai, R. Facchini, A. Giombetti, S. Granchi, A. Ricci, M. Scabia, Ultrasonic images of tissue local power spectrum by means of wavelet packets for prostate cancer detection, in: Proc. 26th Int. Symp. Acoust. Imaging, vol. 26, 2002, pp. 97–104.

28. E. Biagi, S. Granchi, L. Breschi, E. Magrini, F. Di Lorenzo, L. Masotti, Tissue differentiation based on radiofrequency echographic signal local spectral content, in: Proc. IEEE Ultrasonics Symp., Honolulu, USA Oct. 5–8, 2003, pp. 1030–1033.

29. E. Biagi, L. Breschi, S. Granchi, L. Masotti, Metodo e dispositivo perfezionati per l'analisi spettrale locale di un segnale ecografico, Italian Patent FI2003A000254, Oct. 8, 2003.

30. L. Masotti, E. Biagi, S. Granchi, L. Breschi, Method and Device for Spectral Analysis of an Echographic Signal, US Patent, Pub. No. US 2003/0167003 A1, Sep. 4, 2003.

31. L. Masotti, E. Biagi, L. Breschi, S. Granchi, F. Di Lorenzo, E. Magrini, Tissue differentiation based on radiofrequency echographic signal local spectral content (RULES: Radiofrequency Ultrasonic Local Estimator), in: Proc. IEEE Ultrasonics Symp., 2003, pp. 1030–1033.

32. L. Masotti, E. Biagi, A. Acquafresca, L. Breschi, F. Di Lorenzo, S. Granchi, R. Facchini, E. Magrini, F. Rindi, M. Scabia, G. Torricelli, Real time images of local ultrasonic spectral parameters for tissue differentiation through Wavelet Transform, in: Proc. 27th Int. Symp. Acoust. Imaging, vol. 27, 2003, pp. 485–491.

33. L. Masotti, E. Biagi, S. Granchi, L. Breschi, E. Magrini, F. Di Lorenzo, Clinical test of RULES (RULES: radiofrequency ultrasonic local estimators), in: Proc. IEEE Ultrasonics Symp., vol. 3, 2004, pp. 2173–2176.

34. L. Masotti, E. Biagi, S. Granchi, A. Luddi, L. Breschi, R. Facchini, Carotid plaque tissue differentiation based on radiofrequency echographic signal local spectral content (RULES: Radiofrequency Ultrasonic Local Estimators), in: P. IEEE Int. Symp. on Biomed. Imaging, 2008, pp. 1051–1054.

35. S. Granchi, E. Vannacci, E. Biagi, L. Masotti, Differentiation of breast lesions by use of HyperSPACE: hyper-spectral analysis for characterization in echography, Ultrasound Med. Biol 41 (7) (2015) 1967–1980.

36. E. Biagi, S. Granchi, E. Vannacci, L. Lucarini, L. Masotti, Tissue characterization in echographic spectral hyperspace: breast pathologies differentiation, in: Proc. IEEE Ultrason. Symp., San Diego, California, 2010, pp. 1388–1391.

37. E. Biagi, S. Granchi, E. Vannacci, L. Masotti, A. Martegani, First multicenter experience in differentiation of breast lesions by HyperSPACE SPectral analysis in echography, in: presented at 97th Scientific Assembly and Annual Meeting of RSNA, Chicago, 27 Nov.– 2 Dec. 2011.

38. N.D. Kim, V. Amin, D. Wilson, G. Rouse, S. Udpa, Texture analysis using multiresolution analysis for ultrasound tissue characterization, in: Rev. of Progress in Quantitative Nondestructive Evaluation, vol. 16, 1997, pp. 1351– 1358.

39. A. Materka, M. Strzelecki, Texture analysis methods – a review, Report of Technical University of Lodz, Institute of Electronics, COST B11 Report, Brussels, 1-33, 1998, available at: .

40. P.D. Wankhade, A review on aspects of texture analysis of images, Int. J Appl. Innov. Eng. Manage. (IJAIEM) 3 (10) (2014) 229–232.

41. H. Chen, M.J. Zuo, X. Wang, M.R. Hoseini, An adaptive Morlet wavelet filter for time-of-flight estimation in ultrasonic damage assessment, Measurement 43 (4) (2010) 570–585.

42. J. Choi, J.W. Hong, Characterization of wavelet coefficients for ultrasonic signals, J. Appl. Phys. 107 (114909) (2010) 2–6.

43. S.S. Osofsky, Calculation of transient sinusoidal signal amplitudes using the Morlet wavelet, IEEE Trans. Signal Process. 47 (12) (1999) 3426–3428.

44. L. Yue, L. Zhang, Application of Morlet wavelet filter to frequency response functions preprocessing, in: Proc. SEM IMAC-XXII: Conference & Exposition on Structural Dynamics, 2004.

45. J.B. MacQueen, Some methods for classification and analysis of multivariate observations, in: Proc. of 5-th Berkeley Symposium on Mathematical Statistics and Probability, University of California Press, Berkeley, 2015, pp. 281–297.

46. G. Cloutier, Z. Qin, Ultrasound backscattering from non-aggregating and aggregating erythrocytes – a review, Biorheology 34 (6) (1997) 443–470.

47. I. Fontaine, M. Bertrand, G. Cloutier, A system-based approach to modeling the ultrasound signal backscattered by red blood cells, Biophys. J. 77 (1999) 2387– 2399.

48. E. Franceschini, B. Metzger, G. Cloutier, Forward problem study of an effective medium model for ultrasound blood

characterization, IEEE Trans. Ultrason. Ferroelectr. Freq. Control 58 (12) (2011) 2668–2679.

49. E. Franceschini, R. Guillermin, Experimental assessment of four ultrasound scattering models for characterizing concentrated tissue-mimicking phantoms, J. Acoust. Soc. Am. 132 (6) (2012) 3737–3747.

50. E. Franceschini, G. Cloutier, Modeling of ultrasound backscattering by aggregating red blood cells, in: J. Mamou, M.L. Oelze (Eds.), Quantitative Ultrasound in Soft Tissues, Springer, 2013, pp. 117–146 (Chapter 6).

51. R.K. Saha, M.C. Koliosa, A simulation study on photoacoustic signals from red blood cells, J. Acoust. Soc. Am. 129 (5) (2011) 2935–2943.

52. E. Biagi, M. Calzolai, L. Capineri, S. Granchi, L. Masotti, M. Scabia, FEMMINA: a fast echographic multi parameter multi imaging novel apparatus, in: Proc. IEEE Ultrasonics Symp., 1999, pp. 739–748.

53. M. Scabia, E. Biagi, L. Masotti, Hardware and software platform for real-time processing and visualization of echographic radiofrequency signals, IEEE Trans. Ultrason. Ferroelectr. Freq. Control 49 (2002) 1444–1452.

54. T. Fawcett, An introduction to ROC analysis, Pattern Recogn. Lett. 27 (8) (2006) 861–874.

Index